STATISTICS FOR THE BEHAVIORAL SCIENCES

A Short Course and Student Manual

Bryan Raudenbush

University Press of America,® Inc.
Dallas · Lanham · Boulder · New York · Oxford

Copyright © 2004 by
University Press of America,® Inc.
4501 Forbes Boulevard
Suite 200
Lanham, Maryland 20706
UPA Acquisitions Department (301) 459-3366

PO Box 317
Oxford
OX2 9RU, UK

Library of Congress Control Number: 2003098288
ISBN 0-7618-2750-1 (paperback : alk. ppr.)

THIS BOOK IS DEDICATED TO THE JESUITS

CONTENTS

LIST OF TABLES

LIST OF FIGURES

PREFACE

This manual has been prepared as a supplement to any textbook that you may be using for a course in statistics which focuses on behavioral sciences issues. Thus, it is well suited for study in areas such as psychology, sociology, criminal justice, and business.

I believe that it will be quite useful to you, particularly in three major areas:

First, most professors would prefer that your time spent in class consist of listening to what they have to say, rather than writing down everything that they have to say. To alleviate some of your writing, this manual has been designed to outline many of the important concepts that would be covered in an introductory behavioral sciences statistics course. Many of the formulas that you will need to know have been included, and there is space to complete sample problems, which are provided. However, there is no better way to learn material than actually going to class, so this manual should not be looked upon as an "easy way out" of class attendance.

Second, printouts of the sample problems are included for those courses which make use of a statistical computer package known as SPSS (Statistical Package for the Social Sciences). In modern statistical analysis, the use of a computer system is commonplace. This manual provides space for your notes on the various functions of SPSS, or whatever statistical package your instructor may be using.

Third, the use of in-class exercises to make sure that everyone understands the important concepts are necessary for learning statistics. Places are provided in the manual for each of these exercises, with additional exercises being provided for each section.

I hope that your personal experience with behavioral statistics is a favorable one, and I wish you the best as you begin your journey. Being able to competently organize, analyze, interpret, and present data are of paramount importance in any field of study.

--Dr. Bryan Raudenbush

MODULE I: INTRODUCTION TO STATISTICS

The behavioral sciences are empirical sciences. What exactly does that mean? An empirical science is one which relies on observations or experiences to verify or disprove certain phenomenon. We often make observations about the world around us, and it is frequently necessary to quantify or qualify our observations so that we can make decisions about the true nature of the phenomenon of interest. Statistical analysis is concerned with characterizing our observations, analyzing our observations, and making decisions based on the data that we collect.

For example, what if you wanted to know whether or not the incoming freshman class at your university had SAT scores that were higher than the incoming freshman class at a rival university. Through the use of statistics, we can find the answer by comparing the scores between the two universities. Perhaps you are more interested in spending your time at the race track, and are curious about the probabilities associated with betting on different greyhounds, horses, or cars. Statistics can tell you the likelihood of either coming home broke or hitting the big payoff. Maybe you are more science minded, and invented a new drug to cure the common cold. You will need to conduct an experiment to see how people react to your new drug in relation to an alternative drug (or different dosages of the drug) versus a placebo. Statistics can let you know if your patients are getting better, if your drug is better than other drugs currently on the market, and even help you project how much you will need to sell the drug for in order to make a profit.

Obviously, statistics plays an important role, not only in the behavioral sciences, but in any field where you need to make judgments. However, this is a manual designed specifically for the behavioral sciences major. As such, our focus will be primarily on the application of statistical methods to characterizing data and answering questions of a sociological, psychological, or behavioral nature.

THE SCIENTIFIC METHOD

In high school, you may have had a course (perhaps Chemistry or Biology) that introduced you to the "Scientific Method." The Scientific Method is a process through which scientists examine a particular problem or question in the hopes of answering a specific question. The Scientific Method typically involves five steps:

Step1: Statement of a testable hypothesis. This is the experimental question that you would like to have answered. For example, you may be interested in whether there is a difference in the average income between physicians and lawyers. A testable hypothesis might be "The average income of physicians is significantly greater than that of lawyers."

Step 2: Development of an experiment. This is where you decide how you will go about getting your information. Staying with our income example, we might decide to send a questionnaire to physicians and lawyers and have them indicate what their yearly income is. This is a very easy experiment, however, frequently they are much more complicated. For example, we might want to know whether Drug A is better than Drug B at alleviating feelings of stress. Our experiment in this case might involve randomly assigning patients to three groups, in which two of the groups receive either Drug A or Drug B and the third group receives a placebo. Further, we may wish to subject these patients to varying levels of stressful situations and ask them to tell us how "stressed out" they feel. Even this example only illustrates a moderately elaborate experiment; the complexity of the experiment you wish to conduct is limited only by your imagination.

Step 3: Data collection. As the name implies, this is where we collect data on the question(s) of interest. As noted above, this may be income figures for the doctors and lawyers, or measures of stress for our drug groups. The data may take the form of quantitative information (for which a number represents some aspect of the situation) or qualitative data (for which we describe, in words, the current state of events).

<u>Step 4: Data analysis</u>. This is where statistics come into play. Throughout your current course, you will learn which statistical test(s) are appropriate depending on what type of data you have. Once you have performed the statistical tests, you can move on to the final process of the Scientific Method, in which we use these data to draw some conclusions.

<u>Step 5: Draw some conclusions</u>. Our conclusions will be based on the outcomes of our data analysis, therefore it is extremely important to take great care in performing the analyses properly. However, it should be noted that each aspect of the Scientific Method is equally important. If your experimental design is flawed, the conclusions that you draw may be flawed. Likewise, if you collect data in an improper fashion or perform an inappropriate statistical analysis, the conclusions you draw may be worthless.

Although all aspects of the Scientific Method need to be carefully addressed, the focus of this manual will be on steps 4 and 5. Steps 1 through 3 will already be provided for you, and will most likely be further addressed during a course on research methods. Your primary tasks throughout this course will be to decide which types of statistical analyses are necessary, perform these analyses, and draw some conclusions based on the findings.

TWO MAIN METHODS IN STATISTICAL ANALYSIS

Imagine that you are a clinical psychologist and are interested in the study of intelligence. As a first step to your studies, you intend to measure the intelligence of a group of 10th grade students. You administer the test, and you get a I.Q. (Intelligence Quotient) score for each of the students. Now what can you do? With statistics, you are afforded two possible ways to characterize or analyze your data: Descriptive Methods and Inferential Methods.

DESCRIPTIVE METHODS

As the name implies, descriptive methods are used to describe your data. Objectives of descriptive statistics include:

●   Describing a set of events (for example, how did you go about gaining your information?)

●   Summarizing numeric information (for example, how many?)

●   Revealing relationships between variables (for example, how does one variable of interest relate to a second variable of interest?)

INFERENTIAL METHODS

These methods involve decision-making under conditions of uncertainty. One of the questions that you can ask using inferential methods is: "Given this summary of data from a representative sample, what is the probability that some theoretical state of affairs is true?"

Inferential methods rely heavily on the concept of normal probability. For example, imagine that you took a coin and flipped it. What is the likelihood (or probability) that the coin will come out heads? The probability associated with heads is 50 percent, as is the probability associated with tails. This is an easy one. What if you were interested in food preferences between males and females. Let's assume that you gave a list of 500 foods to males and females and you asked your participants to indicate how many of these 500 foods they have tried. When you examined your data, you found that males, on average, had tried 350 of the foods, whereas females had tried 325 of the foods. Obviously there is a numerical difference between the groups, but is this a statistically significant difference? That means, if you repeated this experiment again and again and again, with different samples of males and females, would you consistently find that males had tried more foods than females? That is an inferential question that relies on the concepts of probability. We will address this question later, but without inferential statistics, we would not be able to answer it.

Another major component of inferential statistics is presented in the following equation:

$$\text{Effect} = \text{Treatment} + \text{Noise}$$

In statistics, we seek to analyze mediating effects, such as "The effect of Drug A on athletic performance," or "The effect of television viewing on aggressive behavior." Every time we perform an experiment and find (or do not find) a difference based on some variable of interest (like Drug A or television viewing), we know that this difference is due to two things: Treatment and Noise. Treatment can be thought of as what we, as experimenters, are doing. So, if we give Drug A to a group of athletes, that is their treatment; if we subject people to a certain number of hours of television, that is their treatment. Noise is all the other stuff that can get in the way. For example, imagine we give Drug A to a group of athletes (treatment) and we want to see how well they perform on a treadmill (effect). What are some of the things that may influence their performance (effect) that have nothing to do with Drug A (treatment)? There are dozens: gender, level of fitness, training regimen, motivation, mood, injuries, etc. These factors are all noise, and are mucking up our ability to determine a true effect. Although we can never eliminate noise from statistical analyses, inferential statistics will help us to control some of these factors, and thus allow us to make better conclusions.

ACTIVITY 1.1: Let's return to our example of I.Q. scores. If we were to use descriptive statistics, what types of things might we be able to say?

What about if we were to use inferential statistics?

BASIC TERMINOLOGY

POPULATION AND SAMPLE

- A <u>population</u> consists of all members of some specified group

- A <u>sample</u> is a subset of a population

The differences between a population and a sample can sometimes become muddied depending on the question that you are trying to ask and answer. For example, assume that you are interested in high school SAT scores. The first question that you could ask is: "Are the SAT scores of males and females different at University X?" In this case, your population is all students at University X, because that is the group in which you are most interested. While it might be difficult to obtain the SAT scores of all the students at University X, you could take a sample of 100 males and 100 females at University X in hopes of answering your question.

Let's look at the question in another way: "What is the average SAT score for all people who have taken the SAT?" Here, our population would be the scores of all people who have taken the SAT. Again, it is probably impractical, or impossible, to obtain the SAT scores for the thousands of students who have taken the SAT, so perhaps you just look at the SAT scores of all students at University X. In this case, all University X students would be the sample.

ACTIVITY 1.2: Consider the following research questions, and decide what would be the sample and what would be the population.

1. The manager of a large home improvement warehouse wishes to determine the percentage of shoppers in his store who use a charge card for purchases. He records whether purchases are made with a charge card or cash for each of 100 customers.

Sample:

Population:

2. A survey of 50 families selected from small towns showed that the average number of times per week the families ate the evening meal outside the home was 1.20.

Sample:

Population:

3. One hundred registered voters are questioned by a random phone interview as to their preference of presidential candidates.

Sample:

Population:

INDEPENDENT AND DEPENDENT VARIABLES

When conducting a research study, there are two main variables that need to be identified, which are the independent and dependent variables.

- The <u>independent variable</u> is the variable that an experimenter manipulates or controls in an experiment. It may also be a variable through which the experimenter groups participants.

● The <u>dependent variable</u> is the variable that the experimenter believes will be effected by the independent variable, and is measured in some way.

The dependent variable varies as a function of the independent variable. Typically, the relationship between the two variables can be written as follows:

The effects of _____ (the independent variable) on _____ (the dependent variable).

As an example, let's say we are interested in whether or not relationship status has an effect on how happy you are. We get a sample of single, married, divorced, separated, and widowed participants, and we ask them to tell us how happy they are on a scale from 1-10, where 1=completely happy and 10=completely unhappy.

Our independent variable will be relationship status, because that is how we are grouping our participants. Our dependent variable will be level of happiness, which we will measure, because we believe that relationship status will have some effect on how happy you are.

You will notice that our independent variable (relationship status) has several "levels." A level is an indication of a sub-division of our independent variable. In this case, our independent variable has 5 levels: single, married, divorced, separated, and widowed.

ACTIVITY 1.3: Suppose we were interested in the effects of television viewing on aggression. How could we design a study to assess this effect?

Independent Variable:

Levels of the Independent Variable:

Dependent Variable:

How might we go about measuring the Dependent Variable:

Suppose we were interested in the effects of illumination on worker output in a furniture factory. How could we design a study to assess this effect?

Independent Variable:

Levels of the Independent Variable:

Dependent Variable:

How might we go about measuring the Dependent Variable:

DISCRETE AND CONTINUOUS VARIABLES

- Discrete variables will take on a limited number of different values.

- Continuous variables will take on an infinite number of different values.

An example of a discrete variable is sex. You can be male or female, thereby having only two possible values. Previously, when we assessed happiness as it related to relationship status, both happiness and relationship status were discrete variables. Happiness could be a number from 1-10 and relationship status could be single, married, divorced, separated, or widowed.

An example of a continuous variable would be age. Theoretically, if we never died, we could increase in age forever. Height and weight are also continuous variables, since they are able to increase infinitely.

ACTIVITY 1.4: Note whether the following variables are discrete or continuous, and why:

a.   Models of Chevrolet trucks

b.   Temperature

c.   Class status in college (freshman, sophomore, etc.)

d.   Number of children in a household

QUANTITATIVE AND QUALITATIVE VARIABLES

- Quantitative variables are those variables which have a meaningful number attached to them which represent some intrinsic quantity.

- Qualitative variables are those variables which are represented by some category or attribute.

Let's imagine my dog Cooper. To make it easier for you, Cooper is a black and grey terrier/schnauzer mix with a love of bacon and a penchant for mischief. Quantitative variables that represent Cooper would be her height (18 inches), her weight (25 pounds) and the number of legs she has (four). Qualitative variables that represent Cooper would be her color (black/grey), her breed (terrier/schnauzer) and the texture of her hair (soft).

ACTIVITY 1.5: If you are lucky enough to be near a window, look outside and find a tree. How could you characterize the tree in terms of...

Quantitative variables:

Qualitative variables:

MEASUREMENT

Statistics are calculated from data representing measures of attributes or characteristics. We have just examined some of the forms that these measures may take. Behavioral scientists have been pioneers in measurement theory (for example, clinical assessment, intelligence, attitudes, moods, motivation, etc.).

ACTIVITY 1.6: What are some typical modern day measurement scales that you can think of?

RELIABILITY AND VALIDITY

- **Reliability** relates to the repeatability of a particular measurement.

- **Validity** relates to the ability of a certain measure to actually measure what it is intended to measure.

Reliability is very important in statistical analysis. What if one day you came into the laboratory and we measured your weight to be 150 pounds. Then, 10 seconds later, we measured your weight again and found it to be 180 pounds. Most likely you did not just gain 30 pounds between the first and second measurements. This is an indication that our scale is not reliable. A reliable measurement is one that will produce the same result over and over again.

Validity is a little harder to assess. Suppose you told me that you ate no meat products. I might assume that you didn't like the taste of meat. Is that valid? Maybe or maybe not. Perhaps you don't eat meat because you are a vegetarian, but you do like the taste of meat. If in fact you are a vegetarian and like the taste of meat, and I say that based on your lack of eating meat that you don't like the taste, I am making in invalid statement. Valid measurements or statements are those that accurately address what they are intended to address. To think of this another way, imagine that I measure your height and find you to be 68 inches. I then say that because your height is 68 inches, your I.Q. is 68 points. This would be totally false because the measurement of height is not a valid indication of intelligence.

THE NOIR SYSTEM

Measurement involves assigning symbols to events such that differences among events are reflected in differences among symbols. For our purposes, these symbols are usually numbers. Thanks to S. S. Stevens, a pioneer in measurement theory, we have what is known as the NOIR System of Measurement. NOIR stands for nominal, ordinal, interval and ratio. These are the four major types of measurements with which we will be dealing.

Nominal

Nominal measurements make no assumption of any order or

dimension of the variable of interest. We are simply naming something, and they are therefore qualitative measurements. Sex of a participant is a nominal measurement, since we are just naming people as males or females. Size can also be a nominal measurement if we place objects into categories such as small, medium and large. Your social security number or the number on a soccer jersey are also nominal, since they merely represent a particular object without order or dimension. Statistical analyses are not performed on nominal measurements. After all, we could gain no meaningful information by knowing the average social security number of students in a particular class.

ACTIVITY 1.7: What other nominal measurements can you think of?

Ordinal

Ordinal measurements order events along some dimension, and are therefore quantitative in nature. An example of ordinal measurement would be places in a race: 1st, 2nd, 3rd, and so on. Another example would be your year in college: Freshman, Sophomore, Junior, and Senior.

ACTIVITY 1.8: What other ordinal measurements can you think of?

Interval

Interval measurements have the properties of an ordinal measurement, but they also have a metric. That is, we can measure the intervals between events. We cannot, however, say that one event is a specific proportion greater than or less than another event. Think about temperature. We know that 80 degrees is warmer than 40 degrees, therefore we have an ordinal property. We also know that there is a difference of 40 degrees between the two. Therefore we have an interval. We can not say, however, that 80 degrees is twice as hot at 40 degrees, just like we can not say that 2 degrees is twice as hot as 1 degree.

ACTIVITY 1.9: What other interval measurements can you think of?

Ratio

Ratio measurements have all the properties of an interval scale, plus they have a true zero point which permits ratio comparisons. Height would be a ratio measurement since there is a set zero point (lack of height) and we can make ratio statements, such as 12 inches is twice as long as 6 inches. Money can also be expressed in ratio statements, such as $10.00 is twice as much as $5.00.

ACTIVITY 1.10: What other ratio measurements can you think of?

ADDITIONAL ACTIVITIES

1.   What is the process of the Scientific Method?

2.   Give an experimental example to illustrate the difference between a sample and a population.

3.   In the following scenario, identify the independent and dependent variables, and the levels of the independent variable:

   A physiological psychologist was interested in how well several different medications were at minimizing the symptoms of Obsessive Compulsive Disorder (OCD). She measured the number of obsessive/compulsive episodes in three groups of patients. One group of 15 patients was given the drug Selenicane, one group of 20 patients was given the drug Melandronix, and one group of 18 patients was given the drug Fendipoxide.

4.   In the following scenario, identify the independent and dependent variables, and the levels of the independent variable:

   An experimental psychologist randomly assigned 30 rats to three treatment groups so that there would be 10 rats per group. Only male rats 10 to 12 months old were used. All rats ran the same maze, but the three groups received different kinds of rewards at the end. Group 1 rats found cheese at the end; Group 2 rats found water at the end; Group 3 rats found a female rat at the end. Running speed through the maze was measured for each rat.

5.   Give an example to illustrate both a discrete and a continuous variable.

6.   Identify the following variables as either quantitative or qualitative.

   a.   kinds of drugs
   b.   colors of coffee cups
   c.   weight of a sack of potatoes
   d.   sex of participants
   e.   scores on a statistics exam
   f.   city names

7.   Consider the typical college student.   There are several
        measurements we could make from this student that would
        characterize his or her physical attributes.   What are two
        quantitative and two qualitative measurements that we could make?

8.   Give an experimental example that illustrates good reliability and
        a different experimental example that illustrates poor validity.

9.   Identify the following as one of the four scales of measurement.

        a.   social security numbers
        b.   numbers that represent the major automobile
                manufacturers in the world
        c.   wind speed
        d.   a completely new temperature scale on which 0 is the
                temperature at which Champagne freezes and 100 is the
                temperature of a flaming shot of Bourbon
        e.   birth order
        f.   telephone number
        g.   the value attached to money
        h.   probability statements about the likelihood of rain
        i.   chapters in a book

MODULE II: ORGANIZATION OF DATA

Many times when we perform statistical analyses, we will not be the only ones who need to know the outcome. It is often the case that we will be required to present the data to others, such as journal editors, research colleagues, or the companies who employ us. Therefore, it is very important to understand how to best present our data for maximum efficiency, understanding, and impact.

There are several ways to examine a set of data and/or to present them to an audience. The analyst can...

- <u>Provide all the data</u>. This is fine when you have a small data set, but it is not uncommon for experimental studies to have dozens of variables with hundreds of participants, which would make it almost impossible to derive any meaningful information from this approach.

- <u>Provide numerical summaries of the data</u>. This approach may include presenting information that concisely summarizes the main trends of the data, such as percentages, averages, or frequencies.

- <u>Provide graphical summaries of the data</u>. This approach may also include summarizing the main trends of the data, but in graphical form.

NUMERICAL REPRESENTATIONS

Suppose we have the following information in Table 2.1 concerning the sex of 10 research study participants:

Table 2.1: Sex of research study participants.

Participant	Sex
1	Female
2	Female
3	Male
4	Male
5	Male
6	Female
7	Male
8	Male
9	Male
10	Male

One of the first tasks as a statistical analyst is to look at the data. In addition to whatever else we can learn, this task is particularly important when we are using a computer for data analysis, because we need to ensure that the data have been entered properly. Statistical computer software packages have definitely made the statistician's job easier, however, the analyses are only as good as the data that are input. If an error is made during input, any conclusions we would draw would be incorrect.

If we simply provide the above data to the reader, we have lost no information. However, such a method provides no summary to aid the reader in discerning trends in the data and it reveals nothing about the aggregate of participants. Therefore, we typically choose a more meaningful approach.

FREQUENCY DISTRIBUTIONS

Frequency distributions can be used to summarize the set of data without having to present every single data point. Take note of Table 2.2.

Table 2.2: **Frequency Distribution for Sex of Participants.**

Sex	f	%	rf
Male	7	70.0	0.70
Female	3	30.0	0.30

n=10

In Table 2.2, we are able to get a much clearer view of our sample, without the loss of any information. When producing a frequency distribution table, there are certain elements that should exist. These are the following:

●   Title. The title should clearly state what your table is presenting. Be clear and concise.

●   Column Headings. Column headings tell us what each column represents, such as sex of the participant, the frequency count, the percentage, and the relative frequency.

●   Response Codes. These are the possible values that our data could take. In our example, participants could either be male or female.

●   Sample Size Indication. Sample size is the number of observations that you have taken in your study, or an indication of how many participants completed the task. If we have sample data, we typically denote sample size with a small letter 'n.' If we have population data, we typically denote sample size with a large letter 'N.'

●   Frequency Counts. This is the number of occurrences for our response codes. In this example, there were seven males and three females. Frequency is denoted by a small letter 'f.'

● **Relative Frequencies or Proportions.** Relative frequencies are the proportion of a particular response code in relation to the overall sample size. Since we had seven males and a total of ten participants, the relative frequency of males was 0.70. Relative frequency is denoted as 'rf' and is computed as follows:

$$rf = \frac{f}{n}$$

● **Percentages.** Percentages are another way to show a particular response code in relation to the overall sample size. Percentages are computed as follows:

$$\% = rf \times 100$$

ACTIVITY 2.1: Consider the following data on class standings, where Fr=Freshman, So=Sophomore, Ju=Junior and Se=Senior.

Participant	Class Standing	Participant	Class Standing
1	Fr	11	Fr
2	Jr	12	Se
3	So	13	So
4	Se	14	Se
5	Fr	15	Jr
6	Jr	16	Fr
7	Se	17	Fr
8	So	18	Se
9	So	19	Jr
10	Fr	20	So

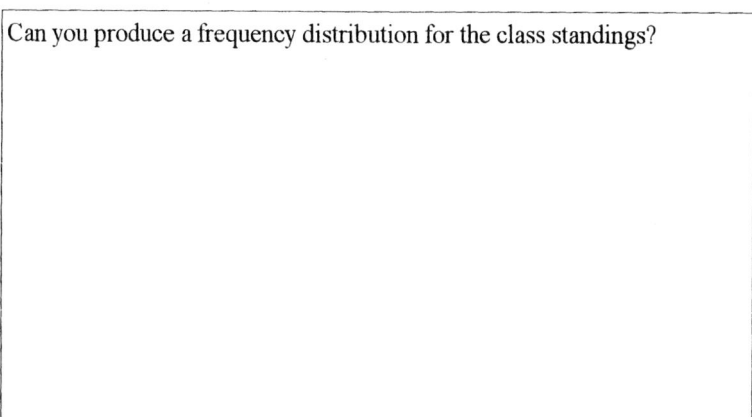

Can you produce a frequency distribution for the class standings?

The above sex of participants example makes use of nominal data. If we have ordinal or higher levels of measurement, our tables can become more elaborate. Take a look at the data in Table 2.3.

Table 2.3: Frequency Distribution for Age of Participants.

Age	f	rf	cf	rcf
25	1	0.01	95	1.00
24	0	0.00	94	0.99
23	2	0.02	94	0.99
22	0	0.00	92	0.97
21	7	0.07	92	0.97
20	15	0.16	85	0.89
19	42	0.44	70	0.74
18	28	0.29	28	0.29

n=95

The example in Table 2.3 introduces the use of two new columns:

● Cumulative Frequencies. This is a cumulative addition of the frequencies. This allows us to know how many people have scores equal to or less than (or equal to or more than, depending on how you arrange your data) a particular score of interest. Response categories must be in order, but whether they are listed from smallest to largest or vice versa is up to you. Cumulative Frequency is denoted as 'cf.' Note that the maximum cf is always = n (or N, if you are dealing with population data).

●   Relative Cumulative Frequencies. This is the same as cumulative frequencies, except with the use of the relative frequencies instead of the actual frequencies. Cumulative percentages can also be used in place of relative cumulative frequencies. Relative Cumulative Frequency is denoted as 'rcf,' and is calculated as

$$rcf = \frac{cf}{n}$$

Note that the maximum rcf is always = 1.00. Also note that response categories with f = 0 should be included for completeness of the table.

ACTIVITY 2.2:   A standardized test consisting of 10 true-false questions is given to a group of 20 people completing a two-hour session in time management. The data of interest are the number of correct answers on the test which is defined as the person's score on the test. The results for the 20 people are as follows:

1	1	3	3	3	3	4	4
5	5	6	7	7	8	8	9
9	10	10	10				

How can we organize a frequency distribution based on the number of correct responses?

GROUPED FREQUENCY DISTRIBUTIONS

You may have too many unique responses to make an ordinary frequency distribution feasible or informative. By grouping response categories together, you can still utilize a frequency distribution. However, it should be noted that there will be some loss of information; for example, you can no longer tell the frequency of one particular score, only the frequency of a group of scores. The number of categories you choose to use depends on personal preference, but 5 to 15 categories is usually acceptable. A grouped distribution is found in Table 2.4.

Table 2.4: Grouped Frequency Distribution of Final Exam Scores in Introductory Psychology Class.

Score	f	rf	cf	crf
95-99	2	0.04	50	1.00
90-94	6	0.12	48	0.96
85-89	10	0.20	42	0.84
80-84	12	0.24	32	0.64
75-79	8	0.16	20	0.40
70-74	6	0.12	12	0.24
65-69	2	0.04	6	0.12
60-64	4	0.08	4	0.08

Note that you should use categories of equal width, and that these categories should be non-overlapping.

STEM AND LEAF DISPLAYS

Stem and leaf displays have the advantage of being both a numerical and graphical way of summarizing data. They also afford you the benefits of a raw data listing in terms of being able to recover the individual data values.

The stem of the display indicates the beginning numbers to a set of data, while the leaf of the display indicates the last number. A stem and leaf display is noted in Table 2.5.

Table 2.5: Example of a Stem and Leaf Display.

Stem	Leaf
6	01334489
5	00011233455677899
4	002556667778
3	55678
2	24
1	8

In the above example, we can tell that the actual scores are 18, 22, 24, 35, 35, 36, and so on. Also, because the line beginning with 5 is the longest, we know that more people scored in the 50's than any other category, while the fewest scores were in the teens.

Note that the shape of a stem and leaf display might be misrepresented if a proportional font were used, so be sure to use a non-proportional font when constructing a stem and leaf display.

ACTIVITY 2.3: Consider the following scores which represent the age at which people retire for a sample of 30 individuals. Can you produce a stem and leaf display?

49	55	56	58	59	60	61	62
62	63	63	63	63	65	65	65
65	65	65	65	65	65	65	66
66	68	70	70	71	71	72	72
72	72	73	73	74	75	80	81

GRAPHICAL REPRESENTATIONS

BAR GRAPHS

Bar graphs are a graphical way of presenting nominal data. The X axis represents your variables or categories, while the Y axis can represent several value types, such as percentages, frequencies, relative frequencies, or averages.

The following figure shows a bar graph of the nominal sex data from Table 2.2. Notice that the bars do not touch each other.

Figure 2.1: Bar Graph for Sex of Participant.

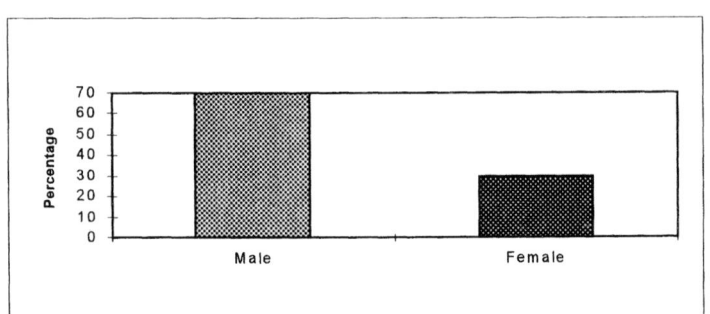

ACTIVITY 2.4: Using the data in Table 2.4, produce a bar graph for letter grades, as follows: A=90-100, B=80-89, C=70-79, D=60-69.

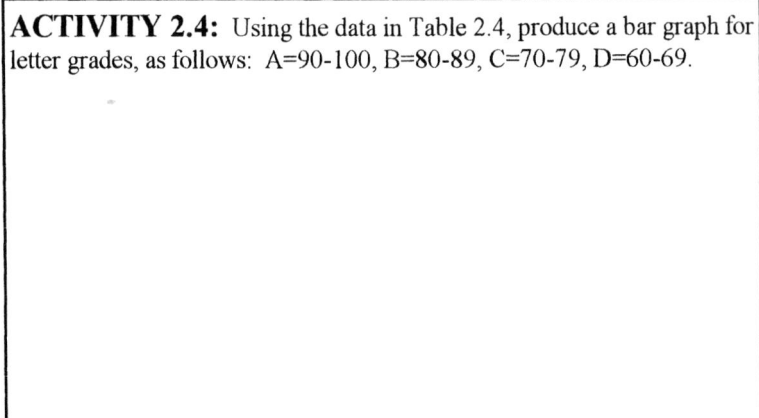

HISTOGRAMS

Histograms are a graphical way of presenting data when our measurement scale is ordinal or higher. Figure 2.2 shows a histogram of the ratio age data from Table 2.3. Notice that the bars touch, indicating that there is a meaningful progression of values. The X axis should always increase from left to right. When using data from a grouped distribution, you could label the X axis using the category limits, the midpoints of the categories, or any other meaningful value.

Figure 2.2: Histogram for Age of Participants.

ACTIVITY 2.5: Using the grouped category data in Table 2.4, produce a histogram.

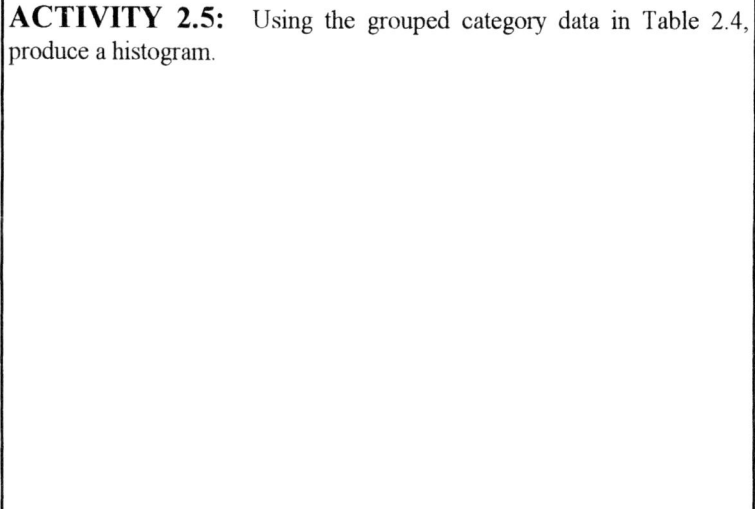

FREQUENCY POLYGONS

The data from Table 2.3 are displayed in a different form in Figure 2.3. This type of graph is known as a frequency polygon.

Figure 2.3: **Frequency Polygon for Age of Participants.**

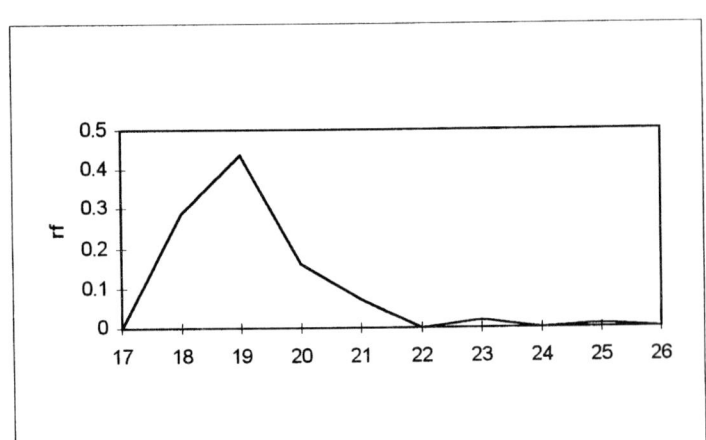

Such graphs can be used with ordinary frequency distributions, although their use is perhaps best reserved for interval or ratio level data. Notice how the bars come down to the X-axis one category below the actual bottom category and one category above the actual top category...age 17 and age 26 are not in our data set, but serve as "anchors" of the polygon line.

ACTIVITY 2.6: Using the grouped category data in Table 2.4, produce a frequency polygon.

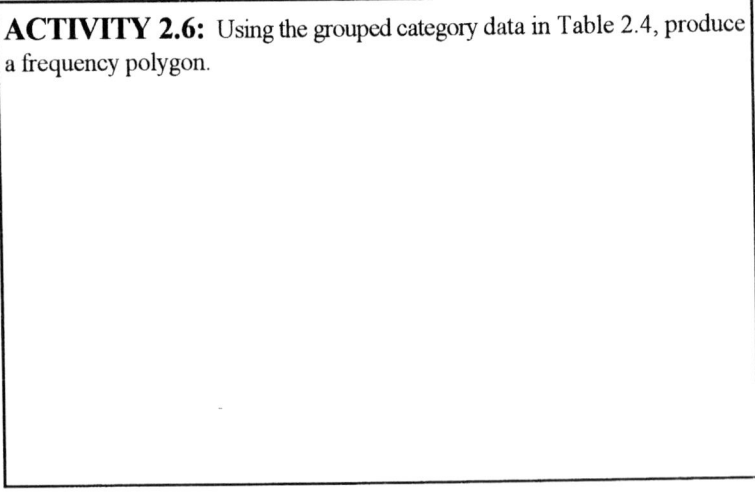

CUMULATIVE FREQUENCY GRAPHS

Figure 2.4 presents the data from Table 2.4 in the form of a cumulative frequency distribution.

Figure 2.4: Cumulative Frequency Graph for Final Exam Scores in Introductory Psychology Class.

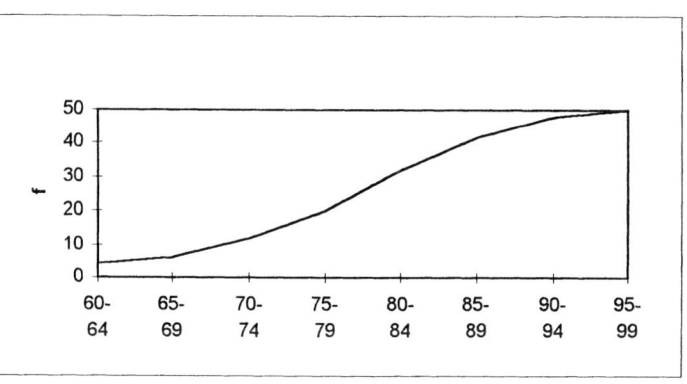

Cumulative frequency graphs increase to some maximum value, such as sample size, percentage, response frequency, etc.

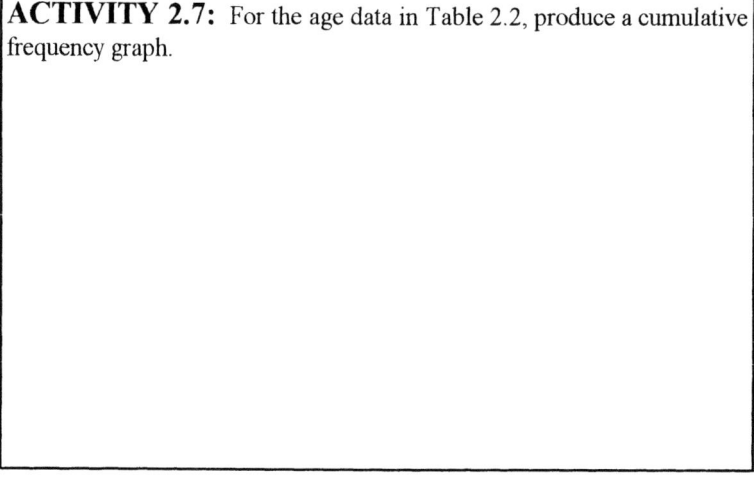

ACTIVITY 2.7: For the age data in Table 2.2, produce a cumulative frequency graph.

PIE CHARTS

While used less frequently than the other types of graphical displays, pie charts are a simple way of presenting nominal data. The entire pie must total 100%, and we can see graphically what percentage each category contributes to the whole. Figure 2.5 is a pie chart of the nominal sex data.

Figure 2.5: Pie Chart of Sex of Participants.

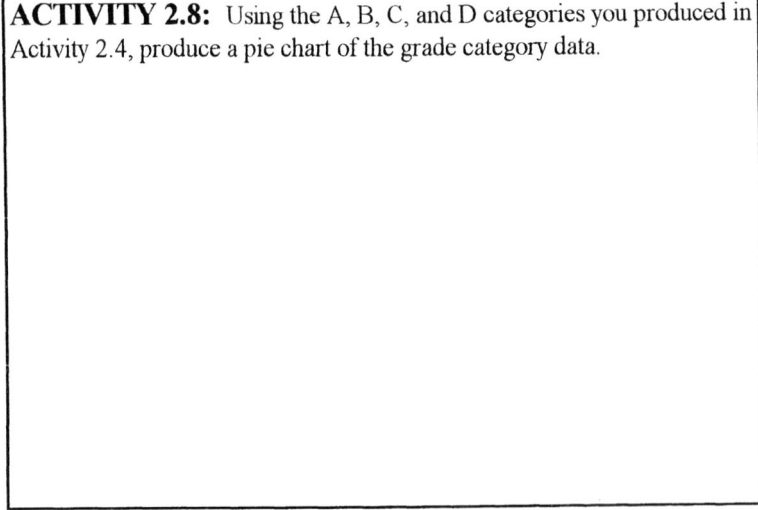

ACTIVITY 2.8: Using the A, B, C, and D categories you produced in Activity 2.4, produce a pie chart of the grade category data.

DESCRIBING DISTRIBUTIONS

When we choose to use graphical displays such as histograms, we are able to notice certain patterns in the data. That is, the data tend to distribute themselves in a certain way based on the frequency of each number's occurrence. There are several major trends in data distributions, which we will now address. Note, however, that in the "real world" most data will only approximate these distribution types.

BELL-SHAPED OR NORMAL DISTRIBUTIONS

Bell-shaped distributions have most of the data in the center of the distribution and the frequencies of occurrence become progressively lower as we near each of the tails of the distribution. Intelligence scores would be an example of a bell-shaped distribution; most people have intelligence scores in the middle range of the scale (average intelligence), while there are few either profoundly dumb or incredibly genius people. Figure 2.6 depicts a bell-shaped distribution..

Figure 2.6: Bell-shaped or Normal Distribution.

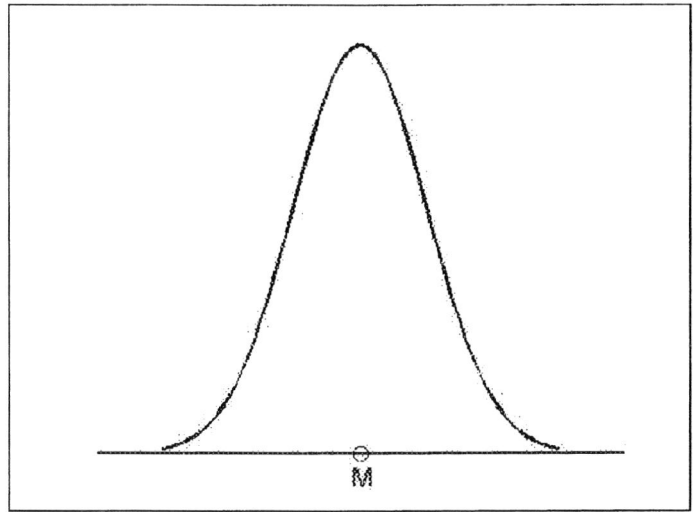

M

ACTIVITY 2.9: What other types of data may distribute themselves in a bell shape?

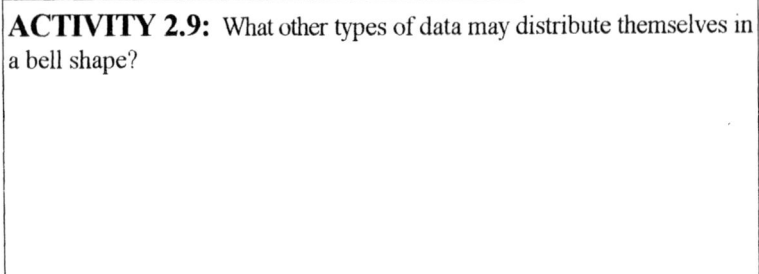

SKEWED DISTRIBUTIONS

Skewed distributions have most of the data on either the right (negative skew) or left (positive skew) of the distribution, and the frequency of occurrence becomes progressively lower as we near the opposite tail. Age at retirement would be an example of a negative skew, since there are very few young people who retire, but many people in the 50-70 year bracket. Age at which people obtain a driver's license would be an example of a positive skew, where most people obtain their license around age 16.

A positively skewed data set has a mean larger than the median, whereas a negatively skewed data set has a mean smaller than the median (these are measures of central tendency which will be discussed in Module III). Negative and positively skewed distributions can be found in Figures 2.7 and 2.8, respectively.

Figure 2.7: Negatively Skewed Distribution.

Figure 2.8: Positively Skewed Distribution.

ACTIVITY 2.10: What other types of data may distribute themselves in a negative skew?

What other types of data may distribute themselves in a positive skew?

SQUARE OR RECTANGULAR DISTRIBUTIONS

Square or rectangular distributions are not noted too frequently in the behavioral sciences, but they play a large role in probability. Imagine that you had a six-sided die and you rolled it 100 times. Most likely, you would have close to an equal number of appearances of 1, 2, 3, 4, 5, and 6. If we were to plot the frequency of each number's occurrence, we would have a rectangular distribution. Likewise, if you flipped a coin 100 times, you would most likely have close to an equal number of heads and tails, thus producing a square distribution. Figure 2.9 depicts a rectangular distribution.

Figure 2.9: Rectangular Distribution.

ACTIVITY 2.11: What other types of data may distribute themselves as a rectangular or square distribution?

BI-MODAL DISTRIBUTION

We may occasionally come across data that present themselves in a bi-modal distribution. These distributions have two peaks in frequency. One example would be worker productivity over time. Some studies have shown that over the course of the day, workers are most productive around 10:00am and 3:00pm, with low points of productivity early in the morning, around lunch time, and late in the day.

Bi-modal distributions are actually a sub-type of what are known as multi-modal distributions. A multi-modal distribution is any distribution that has two or more peaks in frequency. Figure 2.10 shows a bi-modal distribution.

Figure 2.10: **Bi-modal Distribution.**

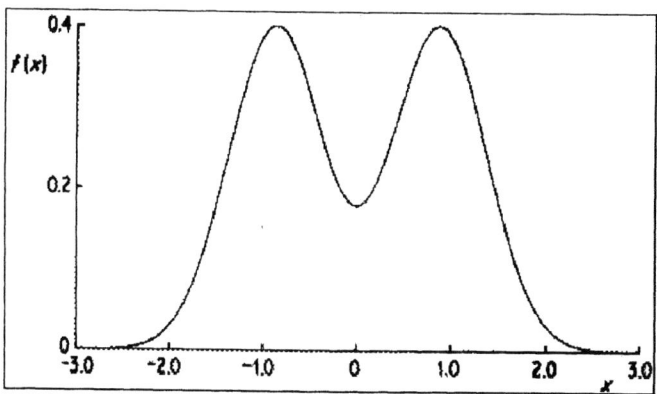

ACTIVITY 2.12: What other types of data may distribute themselves as a bi-modal distribution?

What types of data may distribute themselves as a tri-modal distribution?

ADDITIONAL ACTIVITIES

1.   Draw the following:

    a.   A positively skewed distribution:
    b.   A bimodal distribution:

2.   Consider the following scores:

92	97	94	93	93	93	93
93	96	95	97	93	94	96
97	93	95	94	94	95	93
94	95	94				

    a.   Arrange the scores into a single frequency distribution.
    b.   Draw a frequency polygon from the data.

3.   A charitable organization raised funds to purchase 35 bicycles for needy children. The following table shows the distribution of times (rounded off to the nearest minute) for 35 volunteers to each assemble one of the bicycles.

35	51	57	60	63	64	65
65	65	70	70	72	74	75
75	76	76	77	78	78	79
80	81	82	85	87	88	89
91	93	97	97	97	101	106

    a.   Produce a table of grouped frequency distributions, relative frequencies, cumulative frequencies, and cumulative relative frequencies.
    b.   Produce a bar chart using the grouped frequencies.

MODULE III: MEASURES OF CENTRAL TENDENCY

The three most commonly used measures of central tendency in the behavioral sciences are the mode, the median, and the mean. Each tells us something about the "average" score, or where along some dimension the typical score is located. There are advantages and disadvantages in choosing to characterize your data set with one type of central tendency measure over the other. Therefore, your choice of a measure of central tendency is important.

THE MODE

The mode is the score that occurs most frequently. Referring back to Table 2.1, the modal sex of the participants is male. In Table 2.3, the mode of the age distribution is 19, because more participants reported being 19 years old than any other age. In the grouped frequency distribution shown in Table 2.4, we can see that the modal category is 80-84, but we do not know from the grouped distribution what the modal score is.

A distribution can also be multi-modal, meaning that there is more than one score that occurs with equal greatest frequency. When there are two modes we call this a bi-modal data set, when there are three modes we call this a tri-modal data set, and so on. The mode is abbreviated as "Mo."

In the data set below, the mode is 8 because it occurs most frequently.

X = {4, 6, 6, 7, 7, 8, 8, 8, 8, 9, 10, 10, 12, 15}

In the data set below, we have two modes: 6 and 9. This is a bi-modal distribution.

X = {4, 6, 6, 6, 7, 8, 8, 9, 9, 9, 10, 10, 13, 16}

ACTIVITY 3.1: What is the mode of the following data set?

X = {6, 7, 8, 5, 6, 3, 9, 5, 8, 6, 4, 8, 2, 4}

ADVANTAGES OF THE MODE

The mode can be used with any level of measurement, including nominal, therefore it will work with any type of data you will have. The mode will also be a score (or scores) that occurs in the data set.

DISADVANTAGES OF THE MODE

When we have a very large number of scores, many different values might have approximately equal relative frequencies, thus making the mode less informative. Also, if the mode has been computed from a grouped distribution, it may jump around depending on how the grouping is done. In fact, the "true mode" might not even be contained within the modal category.

THE MEDIAN

The median is the value of a variable that divides the distribution into two equal halves, such that 50 percent of the scores are above the median and 50 percent of the scores are below the median. It will be the score that represents the middle of a distribution. However, that does not imply that the median has to be an actual number in the data set. The median is abbreviated "Me."

Finding the median value is easy when we have an odd number of scores--it is the number that occurs in the middle of the distribution. In the following data set, the median is 7 because it occurs in the middle of the scores.

$X = \{4, 5, 6, 7, 8, 8, 10\}$

When we have an even number of scores, we must take the average of the two middle scores. In the following data set, the median is 12.5. Our two center scores are 12 and 13, making the average of the two 12.5.

$X = \{8, 10, 12, 13, 14, 16\}$

ACTIVITY 3.2:   What is the median of the following distribution of scores?

$X = \{4, 5, 6, 7, 7, 8, 9\}$

What is the median of the following distribution of scores?

$X = \{3, 4, 5, 7, 8, 8\}$

ADVANTAGES OF THE MEDIAN

Often times we would like to be able to talk about the 50th percentile, or the score that separates our data set in half. Therefore, the median is appropriate. The median is also less sensitive to grouping effects than the mode. Finally, the median is insensitive to extreme scores, or outliers, in the data set. Extreme scores or outlyers are scores that are far outside the range of the rest of our scores. We will discuss these soon.

DISADVANTAGES OF THE MEDIAN

Aside from not being able to be used with nominal data, there are few to no other disadvantages.

THE MEAN

The mean is the average of all scores in a distribution. It is computed by adding up all the scores and dividing by N (the total number of scores). This gives rise to the following equation:

$$M = \frac{\Sigma X}{N}$$

"M" stands for the mean. Another common sysbol used for the mean is "$\bar{x}$," and many statistics books use it. We will not. The Greek symbol Σ (upper case sigma) indicates that we should add up what follows the symbol. In this case, X comes after it, which means we should add up all of the individual data points or X points. "X" is a short-hand method of indicating a particular data point, participant, observation, or anything that we can measure.

Consider the following data set...

$$X = \{4, 6, 6, 7, 8, 8, 9\}$$

According to the formula for the mean,

$$M = \frac{(4 + 6 + 6 + 7 + 8 + 8 + 9)}{7} = \frac{48}{7} = 6.8$$

ACTIVITY 3.3: What is the mean of the following data set?

$$X = \{3, 6, 4, 5, 8\}$$

ADVANTAGES OF THE MEAN

The average of all scores in a particular distribution provides meaningful information about our data set. In addition, the mean will play a central role in inferential statistics, as we will see in time.

DISADVANTAGES OF THE MEAN

One of the main disadvantages of the mean is that it is sensitive to extreme scores or outlyers. It can also be influenced by the shape of the distribution, and should be used cautiously with extremely skewed data. Finally, the mean is not typically used with ordinal data, and it would not be useful for nominal data.

INTEGRATION

Which measure of central tendency you decide to use (mean, median, or mode) will depend on the type of data you have and the way you want to convey the data. However, providing only one measure of central tendency will likely not give your reader an adequate understanding of the data set. Thus, when describing a particular data set, all three measures of central tendency should be reported. This is particularly important when you have a data distribution like the one in Activity 3.4.

ACTIVITY 3.4: Imagine that we are interested in knowing something about the "average" income of artists. We send out a questionnaire asking a group of artists to list their yearly income. We get the following data:

| $20,000 | $20,000 | $35,000 |
| $37,000 | $40,000 | $250,000 |

What is the mean?

What is the median?

What is the mode?

Notice that there is a large difference based on whether you choose to use the mean, median, or mode as your measure of central tendency.

ADDITIONAL ACTIVITIES

1. The scores below are from a test of self-esteem, administered to a sample of circus clowns, with larger scores indicating greater self-esteem.

    $$X = \{10, 7, 9, 9, 7\}$$

    Calculate the following...

    a. M
    b. Me
    c. Mo

2. The scores below are from a test of motivation to achieve, administered to a sample of college football players, with larger scores indicating a greater motivation to achieve.

    $$X = \{85, 70, 98, 82, 76, 59, 97, 88\}$$

    Calculate the following...

    a. M
    b. Me
    c. Mo

3. At small colleges and universities there are usually a few courses, typically at the freshman level, that have large enrollments (Introductory Psychology is one such course). However, most of the classes at the junior and senior level have much smaller enrollments. Suppose the mean class size is 23. Would the median class size be larger or smaller, and why?

MODULE IV: MEASURES OF VARIABILITY

In the previous module, we discussed ways of characterizing the central tendencies of our data. It is also important to characterize how spread out the data set is. We will discuss three measures of how spread out the scores in a distribution are: range, variance, and standard deviation.

THE RANGE

The range is simply the difference between the highest score and the lowest score in a distribution. Consider the following data set...

$$X = \{1, 2, 3, 6, 8, 10\}$$

The range would be 9, since 10 - 1 = 9.

ACTIVITY 4.1: From Table 2.3, which lists the ages of several participants, what is the range?

THE VARIANCE

A very important measure of how spread out our scores in a distribution are is called the variance. The variance is based on the sum of the squared deviations from the mean, which we abbreviate as "sum of squares" or "SS." The formula for the sum of squares is as follows:

$$SS = \Sigma(X-M)^2$$

This is the sum of the squared deviations of each score in a data set minus the mean. To derive the sum of squares we must first find the mean (M) of the data set. Then we take each individual score in the data set (each individual "X"), subtract it from the mean, and then square that outcome. Once we have done this for each individual score, we add them up. The result is the sum of squares.

The sum of squares is easy to interpret, however, we are unable to compare the sum of squares between two or more different samples unless the sample size (N) in each comparison sample is the same. To get around that problem, we instead calculate the variance, or the mean squared deviation from the mean.

There is a different formula depending on whether we are calculating the variance of a population or the variance of a sample. Why is that? It is all based on how much we know about a particular data set. When we have population data, we know every score, so we can get a precise measurement of the variance. However, sample data are a sub-set of data from a population, which means that we do not know every score. Therefore, we need to make an adjustment in our variance formula to account for the possibility that our estimate of the sample variance may be somewhat different had we been able to measure every data value in the population.

If we are calculating the variance of a population, we use the following formula:

$$\sigma^2 = \frac{\Sigma(X-M)^2}{N} = \frac{SS}{N}$$

Basically, we are taking the sum of squares and dividing it by the sample size. We use the Greek notation σ^2 (lower case Sigma, squared) to denote a population variance.

If we are calculating the variance of a sample, we use the following formula:

$$S^2 = \frac{\Sigma(X-M)^2}{N-1} = \frac{SS}{N-1}$$

If every score (X) in our distribution has the same value, then the variance will be zero. That means that there is no variability among our scores (they are all the same). The variance increases as the scores become more and more disparate. The variance can never take on a negative value because a squared number can never be negative. Basically, the larger the variance, the larger the variability of the scores.

Table 4.1 presents an example of how to calculate the variance for a group of scores.

Table 4.1: An Example of Calculating the Variance.

X	M	(X-M)	(X-M)²
2	10	-8	64
10	10	0	0
13	10	3	9
15	10	5	25

$$SS=\Sigma(X-M)^2 = 98$$

If these were population data, $\sigma^2 = \frac{98}{4} = 24.5$

If these were sample data, $S^2= \frac{98}{3} = 32.67$

ACTIVITY 4.2: We have gathered data on the number of psychotic episodes for 5 patients at a mental health facility over the course of one month. We found that the number of psychotic episodes were 5, 8, 4, 3, and 5. What is the variance associated with this sample of all patients at the mental health facility?

If these were population data, what would the variance be?

The size and shape of your distribution will also give you some indication about how variable your scores are. Figure 4.1 shows two overlapping distributions, one with high variability and one with low variability. Notice that the more spread out your distribution, the greater the variance.

Figure 4.1: Low and High Variance Distributions.

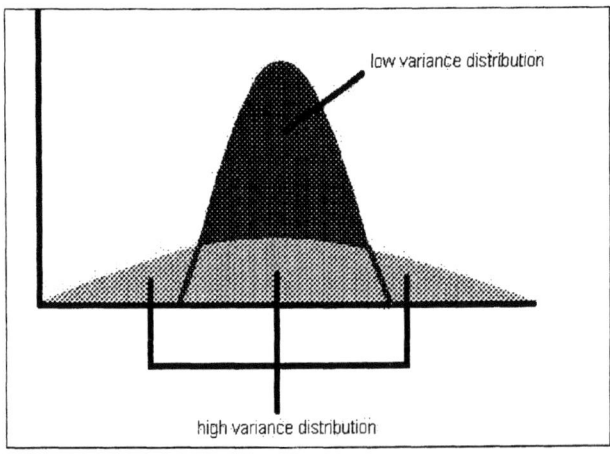

THE STANDARD DEVIATION

Because the variance is expressed in squared units, it becomes difficult for us to relate the variance back to the original units of X. However, by taking the square root of the variance we can obtain a measure of the variability that is expressed in the original units of X. We call this the standard deviation. The standard deviation is denoted as S for a sample and σ for a population, and is calculated as follows:

$$S = \sqrt{S^2} \text{ (for a sample)} \quad \text{or} \quad \sigma = \sqrt{\sigma^2} \text{ (for a population)}$$

Based on the data in Table 4.1, we can calculate the standard deviation for a sample as

$$S = \sqrt{S^2} = \sqrt{24.5} = 4.95$$

and for a population as

$$\sigma = \sqrt{\sigma^2} = \sqrt{32.67} = 5.72$$

ACTIVITY 4.3: In our example of psychotic episodes from Activity 4.2, what would be the variance if we had sample data?

What about if we had population data?

ADDITIONAL ACTIVITIES

1. The scores below are from a test of self-esteem, administered to a sample of circus clowns, with larger scores indicating greater self-esteem.

$$X = \{10, 7, 9, 9, 7\}$$

Calculate the following...

   a.   Range
   b.   SS
   c. S^2
   d.   S

2. The scores below are from a test of motivation to achieve, administered to a sample of college football players, with larger scores indicating a greater motivation to achieve.

$$X = \{85, 70, 98, 82, 76, 59, 97, 88\}$$

Calculate the following...

   a.   Range
   b.   SS
   c. S^2
   d.   S

MODULE V: STANDARDIZED SCORES AND THE NORMAL DISTRIBUTION

An alternative way of measuring the spread of a particular score (X) is in terms of the distance of the score from the mean, expressed in standard deviation units. To do this, we calculate what is called a z score. The formula for a z-score is as follows:

$$z = \frac{(X-M)}{S}$$

A z score affords us the ability to characterize our scores in terms of a) how far away from the mean they are and b) what proportion of scores fell either above or below that score. To really utilize such a technique, however, we need to address the ideas of a standardized distribution, or what has been termed "the normal distribution."

PROPERTIES OF THE NORMAL DISTRIBUTION

Figure 5.1 shows the normal distribution (or normal curve) with a mean of 0 and standard deviations of 1.

Figure 5.1: The Normal Distribution.

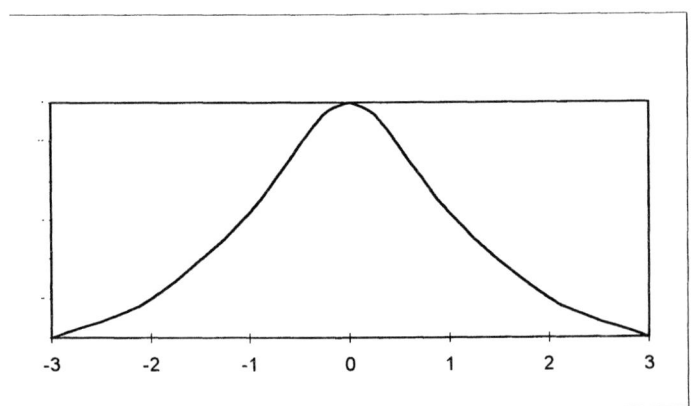

We speak of "the" normal distribution, but in actuality there are an infinite number of normal curves. Each of these curves is symmetrical and range in value from $-\infty$ to $+\infty$.

With the standardized normal curve, we can make use of the laws of normal probability, and designate such probabilities as areas under the normal curve. For example:

- 68% of the time, a variable will take on a value within plus or minus one standard deviation of its mean.

- 95% of the time, a variable will take on a value within plus or minus 2 standard deviations of its mean.

- 99.7% of the time, a variable will take on a value within plus or minus 3 standard deviations of its mean.

These are illustrated in Figure 5.2:

The area under the standard normal curve between the mean (where $z=0$) and a specific value of z (z score) can be found in a table of z-scores (there is one in Appendix A in the back of this manual). In general, since the normal distribution is symmetric and the total area under the normal curve is 1.00 (or 100%), half of the area will lie to the right of the mean and half to the left of the mean. Most z-score tables will give two columns, one indicating the proportion of scores from the mean of the

Figure 5.2: Probability Areas Under the Normal Curve.

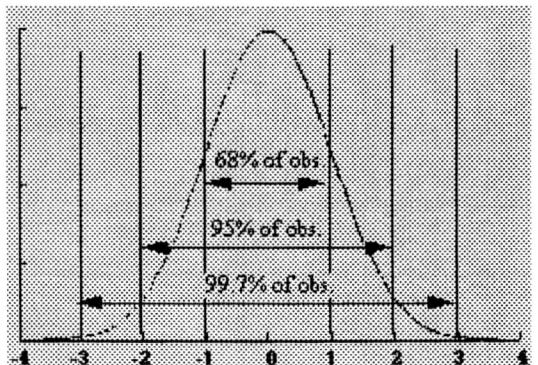

distribution to a particular z score, and one indicating the proportion of scores from a particular z score to the tail end of the distribution. Note that when these two values are summed, you will always get 0.5 (or 50%), which is half of the entire distribution which will always sum to 1.00 (or 100%)

FINDING AREAS UNDER THE NORMAL CURVE

One of the best ways to get a feel for how to derive probability areas under the normal curve is through the working of actual problems. Therefore, let's jump into some.

Let us assume that the mean number of hours of reading done by the average American is 15 hours per week (M = 15), with a standard deviation of 4 hours (S = 4).

QUESTION 1:

What proportion of people read more than 15 hours per week (X = 15)?

$$z = \frac{(X - M)}{S}$$

$$z = \frac{(15 - 15)}{4}$$

$$z = \frac{0}{4} = 0$$

Thus, $p(z > 0) = 0.5$

    This last numerical statement is read as follows: The probability of a z score greater than 0 is .5 (or 50%). This information is derived from the z-score table. You can also say that the proportion of people who read more than 15 hours per week is 50%. As noted in the figure, it is always a good idea to visually indicate the area that you are trying to determine.

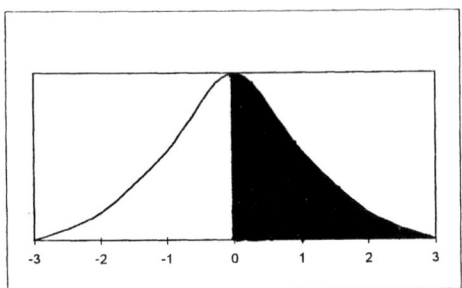

QUESTION 2:

What proportion of people read 20 hours per week or more?

$$z = \frac{(X - M)}{S}$$

$$z = \frac{(20 - 15)}{4}$$

$$z = \frac{5}{4} = 1.25$$

$$p(z > 1.25) = .1056$$

This last numerical statement is read as follows: The probability of a z score greater than 1.25 is .1056 (or 10.56%). You can also say that the proportion of people who read more than 20 hours per week is 10.56%.

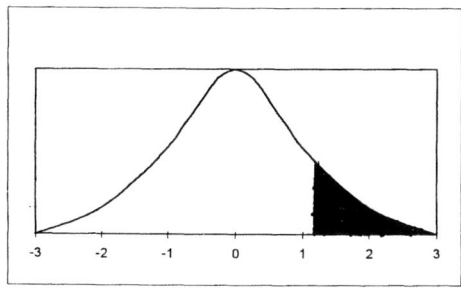

QUESTION 3:

What proportion of people read less than 18 hours per week?

$$z = \frac{(X - M)}{S}$$

$$z = \frac{(18 - 15)}{4}$$

$$z = \frac{3}{4} = .75$$

$$p(z < .75) = .5 + .2734 = .7734$$

Since in this problem we want to find the proportion less than a z of .75, we need to add .5 to the answer to account for the left side of the distribution.

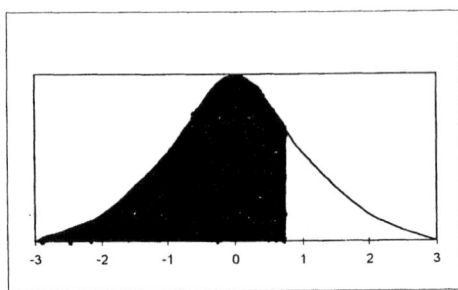

QUESTION 4:

What proportion of people read less than 12 hours per week?

$$z = \frac{(X - M)}{S}$$

$$z = \frac{(12 - 15)}{4}$$

$$z = \frac{-3}{4} = -.75$$

$$p(z<-.75) = .2266$$

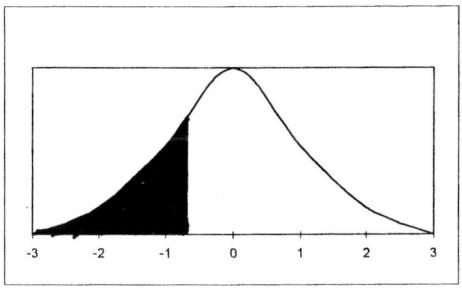

QUESTION 5:

What proportion of people read more than 5 hours per week?

$$z = \frac{(X - M)}{S}$$

$$z = \frac{(5 - 15)}{4}$$

$$z = \frac{-10}{4} = -2.5$$

$$p(z > -2.5) = .4938 + .5 = .9938$$

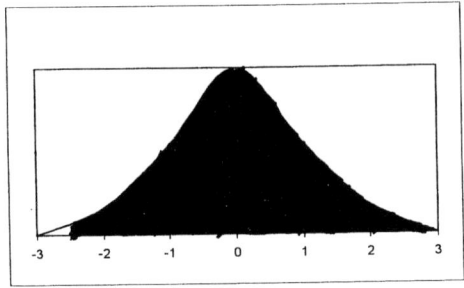

QUESTION 6:

What proportion of people read between 10 and 20 hours per week?

$$z = \frac{(X - M)}{S}$$

$z_{10} = \frac{(10 - 15)}{4} = -1.25$ $\qquad\qquad$ $z_{20} = \frac{(20 - 15)}{4} = 1.25$

$p(-1.25 < z < 1.25) = .3944 + .3944 = .7888$

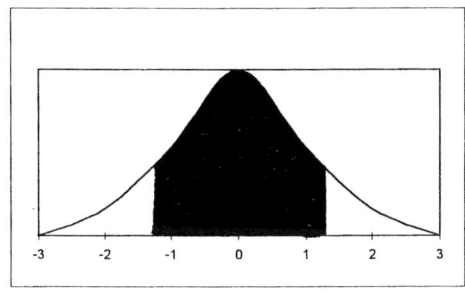

QUESTION 7:

Find the number of hours of reading that separates the middle 95 percent from the lower 2.5 percent and the upper 2.5 percent. For this problem, we need to modify our equation as follows:

$$(z)(S) + M = X$$

Since we want to know the middle 95 percent, that means that each side of the curve will be encompassed by .4750, which corresponds to a z of 1.96.

$$(-1.96)(4) + 15 = 7.16 \qquad\qquad (1.96)(4) + 15 = 22.84$$

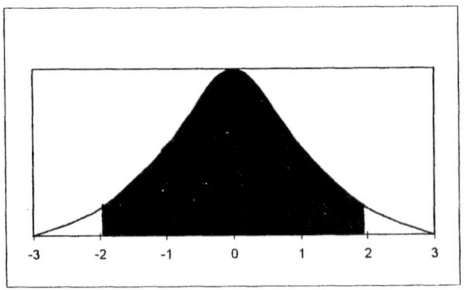

QUESTION 8:

Find the number of hours of reading that encompasses the middle 50 percent.

$(z)(S) + M = X$

Since we want to know the middle 50 percent, that means that each side of the curve will be encompassed by .25, which corresponds to a z of approximately .675.

$(-.675)(4) + 15 = 17.7$ $(.675)(4) + 15 = 12.3$

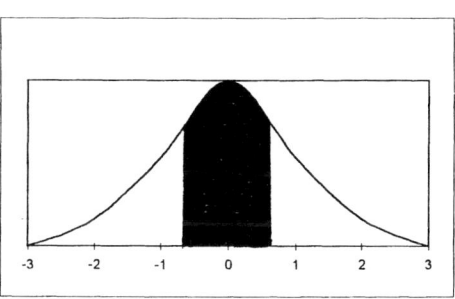

ACTIVITY 5.1: The GRE scores for seniors taking the Psychology Test are normally distributed with a mean of 473 and a standard deviation of 85. What proportion of the seniors would have scores between 450 and 650?

ACTIVITY 5.2: The GRE scores for seniors taking the Psychology Test are normally distributed with a mean of 473 and a standard deviation of 85. What proportion of the seniors would have scores above 700?

ACTIVITY 5.3:   The GRE scores for seniors taking the Psychology Test are normally distributed with a mean of 473 and a standard deviation of 85. If 10,000 psychology majors took the test, how many would be expected to have scores above 670?

ACTIVITY 5.4:   The GRE scores for seniors taking the Psychology Test are normally distributed with a mean of 473 and a standard deviation of 85. Find the scores that separate the middle 95 percent from the lower 2.5 percent and the upper 2.5 percent.

ACTIVITY 5.5: The GRE scores for seniors taking the Psychology Test are normally distributed with a mean of 473 and a standard deviation of 85. Find the scores that encompass the middle 50 percent.

ACTIVITY 5.6: The GRE scores for seniors taking the Psychology Test are normally distributed with a mean of 473 and a standard deviation of 85. The lowest score that a particular Psychology Department allows for admission to the Master's Program is 410. What proportion of the seniors would obtain at least 410?

USING Z-SCORES AS A BASIS FOR COMPARISON

Calculating z-scores is also an excellent way of comparing someone's performance on two different tests, which will have different means and standard deviations. Table 5.1 illustrates this point.

Table 5.1: Comparing Performance on Two Tests Using z-Scores.

	Test 1	**Test 2**
M	80	90
S	15	10
X	85	85
z-score	.33	-.5

In this example, an individual took two different tests, and scored an 85 on each of them. The question becomes, "On which test did she have the better performance?" In order to answer this, we need to compute a z-score for each test. We can see that on Test 1, her z-score is .33, placing her above the mean of the distribution. However, on Test 2 her z-score is -.50, placing her half a standard deviation unit below the mean of the distribution. Based on z-scores, we know that she performed better on Test 1 than Test 2, even though her individual scores were exactly the same. Therefore, even if someone were to receive the same score on two different exams, we can never be sure of exactly where they fall along the distribution without calculating a z-score.

ACTIVITY 5.7: Assume that you took two different tests, one in biology and one in psychology, and you scored a 80 on each exam. The means and standard deviations for the exams are listed below. On which test did you have the better performance? Why?

	Biology	**Psychology**
M	78	85
S	10	5
X	80	80

Sometimes we need to work backwards, and determine the raw score that corresponds to a particular standard score. For example, what score corresponds to a z of 2.0 on Test 1?

$$z = \frac{(X - M)}{S}$$

$$2.0 = \frac{(X - 80)}{15}$$

By cross multiplication we know that...

$$X - 80 = 2.0 \times 15$$

$$X - 80 = 30$$

$$x = 110$$

ACTIVITY 5.8: Based on the information given in Activity 5.7, what is the raw score that would correspond to a z-score of 1.5 on the psychology exam?

ADDITIONAL ACTIVITIES

1.   Consider the following scenario: After conducting an experiment
     to determine the IQ score of 50 12-year-olds, you have found a
     distribution of IQ scores, with a mean of 100 points and a standard
     deviation of 15 points.

     a.   68% of the actual scores received on the test will have a
          range between the scores of _____ points and _____
          points.
     b.   99.7% of the actual scores received on the test will have
          a range between the scores of _____ points and _____
          points.
     c.   50% of the actual scores will be equal to or greater than
          _____points.

2.   Consider the data from the following table. Assuming that a
     student received a score of 130 on both exams, on which exam did
     the student have the better performance, and why?

	Test 1	**Test 2**
M	60	80
S	5	6

3.   Consider the data from the following table. Assuming that a
     student received a score of 80 on both exams, on which exam did
     the student have the better performance, and why?

	Test 1	**Test 2**
M	140	120
S	5	6

4.   Light bulbs produced by a certain company have a mean length of
     life of 200 days with a standard deviation of 15 days.

     a.   What is the probability that a randomly selected light bulb
          will last between 150 days and 200 days?
     b.   What is the probability that a randomly selected light bulb

will last between 120 days and 170 days?

c.   What is the probability that a randomly selected light bulb will last less than 130 days?

d.   What are the numbers of days that encompass the middle 95%?

5.   Suppose a population was normally distributed with a mean of 10 and a standard deviation of 2.

a.   What proportion of the scores would lie between 7.5 and 12.5?

b.   If there were 250 members of the population, how many would be expected to score 11 or more?

c.   What score would separate the lower 40% of the population from the upper 60%?

6.   Consider the following data set: $X = \{2, 7, 3, 6, 2\}$

a.   What is the z score associated with a raw score of 7?

b.   What raw score is associated with a z of 2?

MODULE VI: CORRELATION AND REGRESSION

DESCRIBING RELATIONSHIPS

The statistical methods we have encountered thus far have focused on a single variable (X). Oftentimes, however, we are more interested in how two variables are related to one another. For example, how does I.Q. score relate to GPA? How does weight relate to caloric intake? How does the amount of money you spend on a date relate to how attractive your date thinks you are? We will now shift focus to the problem of describing how two variables are related to one another.

When attempting to describe relationships between variables, our end goal is to answer the following three questions:

1. Are the two variables related?
2. If are two variables are related, what is the nature of the relationship?
3. If our two variables are related, and we can determine the nature of that relationship, can we predict the score of one variable from the score of the other variable?

To answer questions one and two, we will be using a statistical technique called correlation. Correlation analysis allows us to determine whether or not two variables are related, and, if they are related, to determine the nature of that relationship. For example, as the first variable increases in number, does the second variable also increase in number? Or does it decrease in number? How much does the second variable increase or decrease in number for every unit increase or decrease in number of the first variable? To answer question three, we will be using a statistical technique called regression. Regression analysis is used to estimate the value of one variable, given information about the second variable (or, in the case of multiple regression, given information about several other variables). When two variables are being used, the variable that is being predicted is called the dependent variable. The variable being used to predict the value of the dependent variable, that is, the variable for which we already know the value, is called the independent variable. Our discussion will focus on the most basic form of regression analysis: one dependent variable and one independent variable. However, more complicated regression analyses which take into account several independent variables, are also possible and quite typical.

Let's begin with an example. Suppose you are an exercise physiologist, and you are interested in muscular strength. You decide that one of the measures you will take is grip strength, calculated by having someone squeeze a hand grip device (known as a dynamometer) which measures applied force. What other variable might also be related to how much force you are able to exert on the hand grip device? Perhaps muscle size. So, for your second variable, you decide to measure the circumference of the forearm muscle. We now have two variables: grip strength and forearm muscle size.

Listed below in Table 6.1 are the grip strengths and forearm muscle sizes for ten athletes. To make life easier, for further analyses we will simply refer to the two variables as X (hand grip strength) and Y (forearm muscle size).

Notice that since we have more than one variable for which we need to derive a mean, we must distinguish which mean is which. To do this we use subscripts of the variables of interest; for example, M_X is the mean for variable X.

Our next step is to compute some important intermediate statistics, which will later be used when we perform the correlation and regression analyses. These intermediate statistics are shown in Table 6.2.

Note that what we are doing is first calculating how far each X variable is from the mean of X ($X-M_X$) and how far each Y variable is from the mean of Y ($Y-M_Y$). From that point we square each of these deviations from the mean and add them up...you should remember this as the sum of

---

Table 6.1: **Raw Data for Grip Strength and Forearm Muscle Size.**

Athlete	Hand Grip Strength (X)	Forearm Muscle Size (Y)
1	83	34
2	42	23
3	56	23
4	34	24
5	66	29
6	50	26
7	37	27
8	64	34
9	58	23
10	70	27
	$M_X = 56$	$M_Y = 27$

Table 6.2: **Intermediate Calculations for Grip Strength and Forearm Muscle Size.**

Athlete	(X)	$(X-M_X)$	$(X-M_X)^2$	(Y)	$(Y-M_Y)$	$(Y-M_Y)^2$	$(X-M_X)(Y-M_Y)$
1	83	27	729	34	7	49	189
2	42	-14	196	23	-4	16	56
3	56	0	0	23	-4	16	0
4	34	-22	484	24	-3	9	66
5	66	10	100	29	2	4	20
6	50	-6	36	26	-1	1	6
7	37	-19	361	27	0	0	0
8	64	8	64	34	7	49	56
9	58	2	4	23	-4	16	-8
10	70	14	196	27	0	0	0
Σ	560	0	2170	270	0	160	385

---

squares (SS). We will also calculate the product of $(X-M_X)(Y-M_Y)$ and add these up.

Based on these intermediate calculations, we can produce a summary table of important information, as is found in Table 6.3.

Table 6.3: **Summary Table for Grip Strength and Forearm Muscle Size.**

	Hand Grip Strength (X)	**Forearm Muscle Size (Y)**
Mean (M)	56	27
Sum (Σ)	560	270
Sum of Squares (SS)	2170	160
Variance (S^2)	241.11	17.78
Standard Deviation (S)	15.53	4.22

$\Sigma[(X-M_X)(Y-M_Y)] = 385$

Now that we have information on our two variables, we can learn a new type of graphical representation. When the values of two different variables are plotted, the resulting graph is called a scatterplot. The independent variable, usually denoted by X, is plotted along the horizontal axis, and the dependent variable, usually denoted by Y, is plotted along the vertical axis. This has been done for our example in Figure 6.1.

Figure 6.1: **Scatterplot of Grip Strength and Forearm Muscle Size Data.**

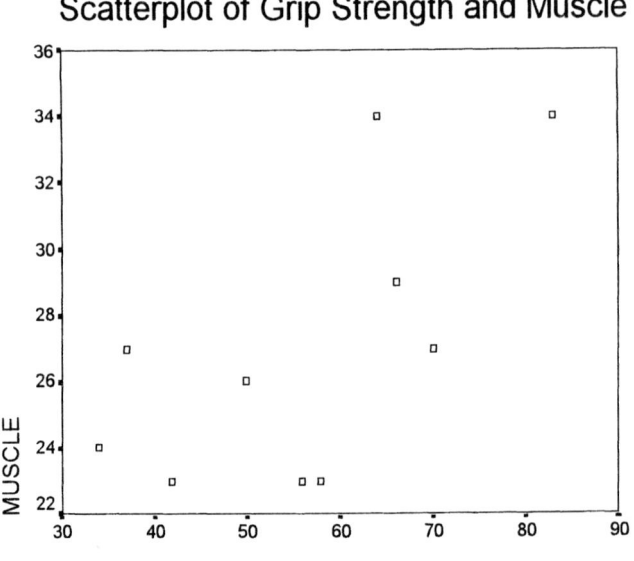

One dot is placed on the graph as a function of both the X and Y variables.

ACTIVITY 6.1: We are interested in the relationship between the number of years a person has been sentenced to prison, and the severity of the crime. We gathered data on how many years each of 10 convicted felons were sentenced to prison, and we had a group of judges rate the severity of the crime on a scale from 1-10, with 1=low severity and 10=high severity. The raw data are below

Prisoner	**Sentence Years (X)**	**CrimeSeverity(Y)**
1	4	3
2	18	8
3	6	3
4	12	7
5	10	5
6	7	5
7	22	9
8	14	8
9	8	4
10	3	3

First, calculate the intermediate statistics

Prisoner	(X)	(X-M_X)	(X-M_X)^2	(Y)	(Y-M_Y)	(Y-M_Y)^2	(X-MX)(Y-MY)
1	4			3			
2	18			8			
3	6			3			
4	12			7			
5	10			5			
6	7			5			
7	22			9			
8	14			8			
9	8			4			
10	3			3			
Σ							

Next, make a table of summary statistics

	Sentence Years (X)	CrimeSeverity(Y)
Mean (M)		
Sum (Σ)		
Sum of Squares (SS)		
Variance (S^2)		
Standard Deviation (S)		

$\Sigma[(X-M_X)(Y-M_Y)] =$

COVARIANCE

The covariance gives us both the direction of the relationship between our two variables (X and Y) and an indication of how strong this relationship is. The covariance (abbreviated as COV) is the mean of the cross-product deviations, and is calculated as follows:

$$COV = \frac{\Sigma[(X-M_x)(Y-M_y)]}{N-1}$$

You may notice that the covariance is very similar in structure to the variance.

In our grip strength and forearm muscle size example, the covariance is calculated as:

$$COV = \frac{385}{9} = 42.78$$

We can use the following rules to interpret the covariance:

●   If the covariance is positive, there is a positive association between X and Y; that is, as the value of X increases, the value of Y increases.

●   If the covariance is negative, there is a negative association between X and Y; that is, as the value of X increases, the value of Y decreases (or as the value of X decreases, the value of Y increases).

●   If the covariance equals zero, there is no association between X and Y.

ACTIVITY 6.2: For the data from Activity 6.1, calculate the covariance for the number of years a person has been sentenced to prison and the severity of the crime.

CORRELATION

The covariance is of limited usefulness as a descriptive measure because its range, while theoretically $\pm \infty$, is a function of the variances of both X and Y, specifically. It is, therefore, hard to compare covariances from one study to another. Remember in the past module what we did to compare two different scores that a person received on two different tests? We had to calculate a z-score. In other words, we needed to standardize the values of each test. We need to do the same in this case. If we convert the covariance to a standardized form, the result is called the Pearson Product Moment Correlation Coefficient (or, more simply, the correlation), symbolized as "r". Once we have calculated the covariance, the correlation is very easy to obtain, as seen below.

$$r = \frac{COV}{S_x S_y}$$

In our example,

$$r = \frac{42.78}{(15.53)(4.22)} = 0.65$$

Correlations range from 0 to 1.00, and can be either positive or negative. The actual number tells us how strong the correlation is, while the sign (either positive or negative) tells us the direction of the effect. We can use the following as a guideline for correlation interpretation:

● If $r = 1.00$, there is a perfect positive association between X and Y.

● If $r > 0$, there is a positive association between X and Y that gets stronger as r gets closer to 1.

● If $r = 0$, there is no relationship between X and Y.

● If $r < 0$, there is a negative association between X and Y that gets stronger as r gets closer to -1.

● If $r = -1.00$, there is a perfect negative association between X and Y.

With practice, you will be able to spot differences in the scatterplot of two variables and be able to estimate how strong the correlation is and in what direction it is (either positive or negative). For example, take a look at the following scatterplots in Figures 6.2, 6.3, and 6.4. They will give you an idea of how your data may distribute themselves in different correlations.

Figure 6.2: **Scatterplot Example of a Strong, Positive Relationship.**

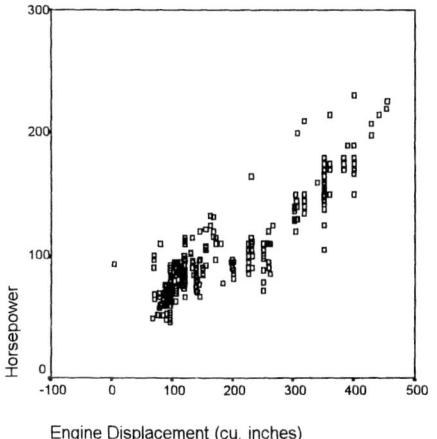

Figure 6.3: **Scatterplot Example of a Near Zero Relationship.**

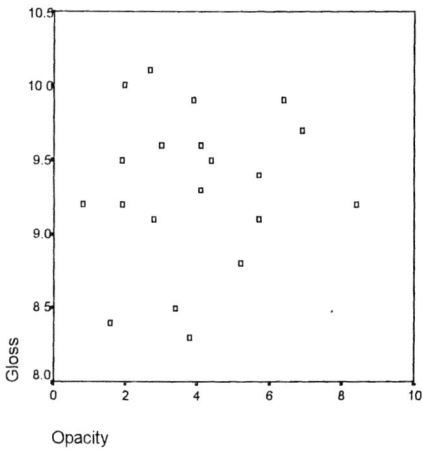

Figure 6.4: **Scatterplot Example of a Moderate, Negative Relationship.**

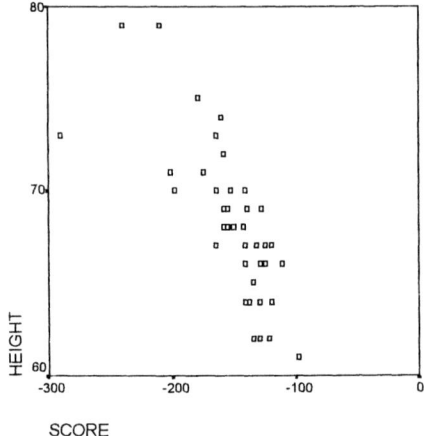

SCORE

So what is a correlation, anyway? A correlation is an indication of how much our two variables overlap in terms of their variances. The more the overlap, the higher the correlation. If the two variables do not overlap at all, the correlation will be 0. If they overlap perfectly, such that a change in one variable leads to a proportional change in the other variable, the correlation will be 1.00. Rarely, if ever, will you find a correlation that equals 1.00.

ACTIVITY 6.3: From the data of Activity 6.1 and 6.2, calculate the correlation between the number of years a person has been sentenced to prison and the severity of the crime.

FACTORS AFFECTING THE CORRELATION

Do you remember our discussion of reliability and validity? There are some things that will affect the reliability and validity of your correlation. That means, there are some conditions under which your correlation coefficient (r) will not adequately represent the true nature of the relationship between your two variables. Some of these conditions are the following:

- **SHAPES OF DISTRIBUTIONS.** If the data sets for your two variables are skewed, particularly if they are each skewed in opposite directions, your correlation may not adequately represent the relationship.

- **SMALL SAMPLE SIZES.** Sample size is directly related to the strength of the relationship. Small sample sizes can drastically affect your correlation.

- **OUTLIERS OR EXTREME SCORES.** Just one data point that is outside the range of the rest of your data points can alter the size of the correlation by 30% or more.

- **RESTRICTION OF RANGE.** Restriction of range involves a large number of observations being nearly equal. Imagine if one of your correlation variables was a measure of how much people like pizza on a scale from 0-10. Most people would indicate a score somewhere between 8-10, which would cause a restriction of range, and thus impact your correlation.

- **NON-LINEARITY.** Correlations look for linear (or straightht line) trends in the data. What if your data are best represented by a shape other than a straight line? You may get a very small correlation, indicating that there is no relation; however, there is a relation, just not a linear relation. For example, performance and arousal data often present themselves in a curvi-linear fashion, such that for low and high levels of arousal, performance is relatively low, but for moderate levels of arousal, performance is high. In this case, you would get a correlation close to 0 (r=0.00), but there is obviously a relationship between the two variables. However, since it is not a linear relationship, the correlation is unable to adequately represent the trends in the data.

PROPORTION OF VARIANCE ACCOUNTED FOR

In our grip strength and forearm muscle size example we found that the correlation was 0.65. How much do our variable variances overlap? To determine this, we calculate what is called the proportion of variance accounted for, or r^2.

To understand the concept of "proportion of variance accounted for" take a look at Figure 6.5.

Figure 6.5: Illustration of the Proportion of Variance Accounted For (r^2).

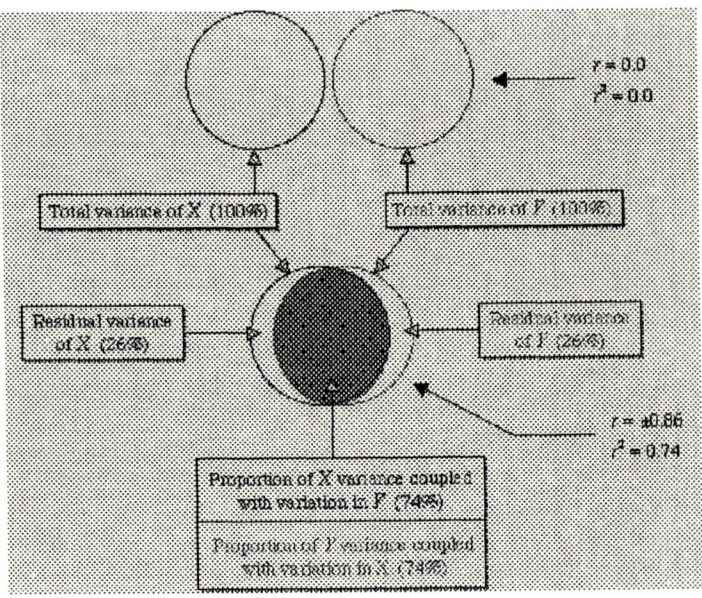

Here we have two variables (X and Y) represented by two circles (this is known as a Venn Diagram). In the top of the figure, these variables do not overlap at all. That means that a change in the X variable is not associated with any discernable change in the Y variable. Thus, r = 0.

In the bottom of the figure we see that there is overlap, and that the correlation (r) is 0.86. Since the proportion of variance is calculated as r^2, the proportion of variance here is 0.74 (calculated as 0.86^2). That means that the variability associated with the two variables overlaps by 74%. So, we are able to account for 74% of the variability in the X variable by knowing the variability of the Y variable. There is still 26% of the variability in each variable that is left unaccounted for. Why is there variability unaccounted for? Mainly because there are usually more things associated with one variable (X) than just one other variable (Y). The variability that we are unable to account for is known as the residual.

In our grip strength and forearm muscle size example, since r=0.65, r^2=.42, or 42%. Therefore, we are able to account for 42% of the variability of grip strength as being associated with forearm muscle size. That means that there is still 58% of the variability unaccounted for (the residual). How could we account for more variability? Perhaps if we chose people who were all the same age, or the same gender, or who were equally motivated to squeeze our hand-grip device. Being able to account for more variability involves eliminating any other factors that may also play a role in accounting for the variable of interest (remember our discussion of "noise" in Module I?). This is why correlations are usually much smaller than 1.00.

One other important point is that "correlation does not imply causation." That is, when we interpret a correlation, we can only say that there is a relationship between the two variables, and not that one variable has caused another variable. For example, imagine that I asked two students in the class (Student X and Student Y) to stand up, and they do. Therefore, there is a perfect positive correlation between the behavior of Student X and the behavior of Student Y (r=1.00, or as one student stood up, the other stood up). However, Student X did not cause the behavior of Student Y, some other factor did...me. So, we can not say anything about causality in correlations, only relationships.

ACTIVITY 6.4: Based on the correlation you found in Activity 6.3, what is the proportion of variance accounted for between the number of years a person has been sentenced to prison and the severity of the crime?

REGRESSION AND THE LINE OF BEST FIT

We have one last question to answer of our trilogy of questions: "Can we predict the value of one variable by knowing the value of the second variable?" Regression analysis will provide us with a straight line to the data in a scatterplot based on a prediction equation. This estimated regression line, or "line of best fit," is obtained through what is called the least squares method. This means that a line will be drawn which (overall) minimizes the distance of each of your data points from the line.

Consider the data and graph in Figure 6.6.

Figure 6.6: Least Squares Method in Regression.

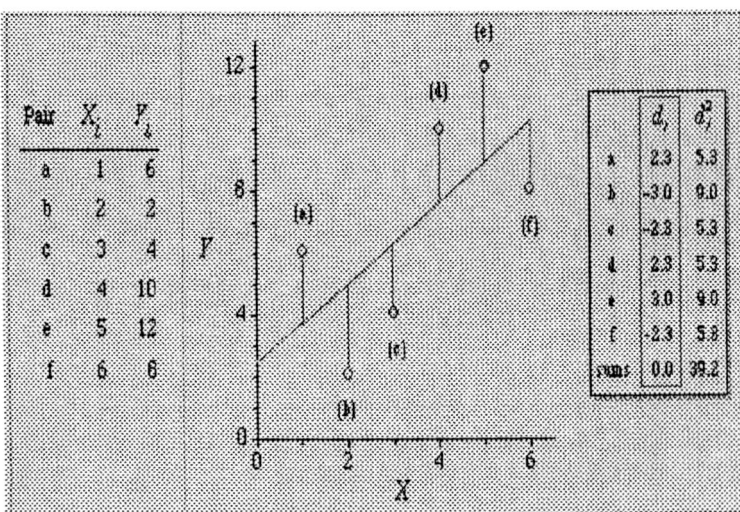

There are six data points (a through f) on a scatter diagram. The regression line (line of best fit) indicates the smallest overall distance possible from each of the data points, such that a straight line can be drawn. Based on this regression line, we can predict one score from another. The equation for the regression line is as follows:

$$Y' = a + bX$$

where "a" refers to the Y-intercept, or the point where the line crosses the

X-axis, and "b" refers to the slope of the line, or the change in Y associated with a unit change in X. For a straight line, the slope is constant.

Notice that we use Y' rather than Y to indicate that the equation produces "predicted" Y values rather than observed Y values. Why predicted? Because we usually perform a regression analysis in order to predict some score in the future. College admission offices do this all the time. These folks have gathered data from past students, such as their SAT scores, high school grades, college grades, educational background, extracurricular activities, etc. The goal of the admissions office is to use this past information in order to predict how well someone will perform in the future, while they are a student at the college. If, once your individual data are entered into the prediction equation, it shows that you will perform well, you are likely to be admitted; if the prediction equation shows that you are likely not to perform well, you may not be admitted. It is important to note, however, that this is only a prediction; there are always exceptions.

COMPUTING a AND b

In computing a prediction equation,

$$a = M_y - bM_x$$

and

$$b = \frac{\Sigma[(X - M_X)(Y - M_Y)]}{\Sigma(X - M_X)^2}$$

An alternative way to compute the slope of the regression line can be made when we already know the correlation and standard deviations of the two variables. When we know these calculations,

$$b = r\,\frac{(S_Y)}{(S_X)}$$

In our grip strength and forearm muscle size example,

$$b = \frac{385}{2170}$$

$$= 0.18$$

Having found b, we can solve for a as follows:

$$a = M_Y - bM_X$$

$$a = 27 - (.18)(56) = 16.92$$

Therefore, our regression line is as follows:

$$Y' = a + bX$$

$$Y' = 16.92 + (.18)(X)$$

In order to place the regression line (or line of best fit) on the scatterplot, we need to insert at least two different values of X into our regression equation in order to obtain at least two different points on the graph through which we can draw the regression line.

Given our regression formula, let's substitute the values of 40 and 60 for X.

$$Y' = 16.92 + (.18)(40) = 24.12$$

$$Y' = 16.92 + (.18)(60) = 27.72$$

We can now draw the regression line through the original scatterplot, as shown in Figure 6.7.

We now have a regression line that will allow us to predict one value from the other. What if someone had a grip strength of 50...could you predict what their muscle size will be? Yes. All you need to do is draw a line up from 50 until you reach the regression line, and then work over to the muscle size variable. Therefore, with a grip strength of 50, you would estimate that their muscle size was 25.

Figure 6.7: **Scatterplot and Regression Line for Grip Strength and Forearm Muscle Size Example**

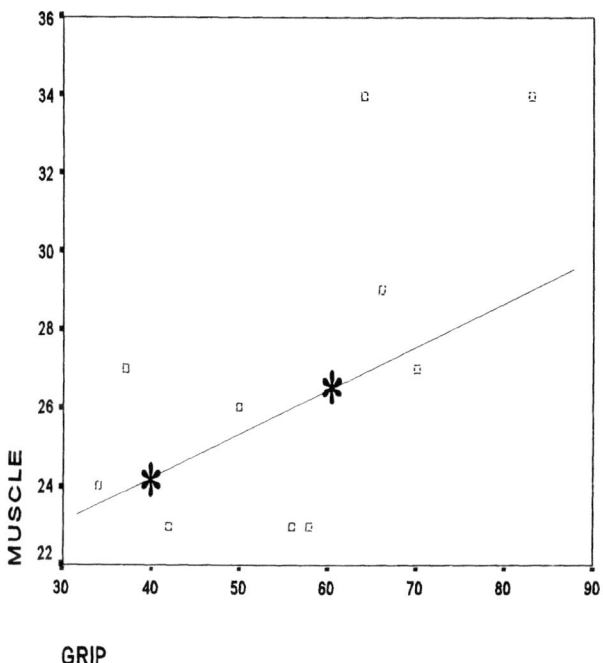

Scatterplot of Grip Strength and Muscle Size

ACTIVITY 6.5: Based on the information you obtained in Activity 6.1, 6.2, 6.3, and 6.4, calculate the regression formula.

Using your regression equation, substitute the numbers 5 and 15 for the number of years sentenced to prison so you can get the associated numbers for the severity of the crime.

Based on these numbers, place your regression line on the scatterplot below.

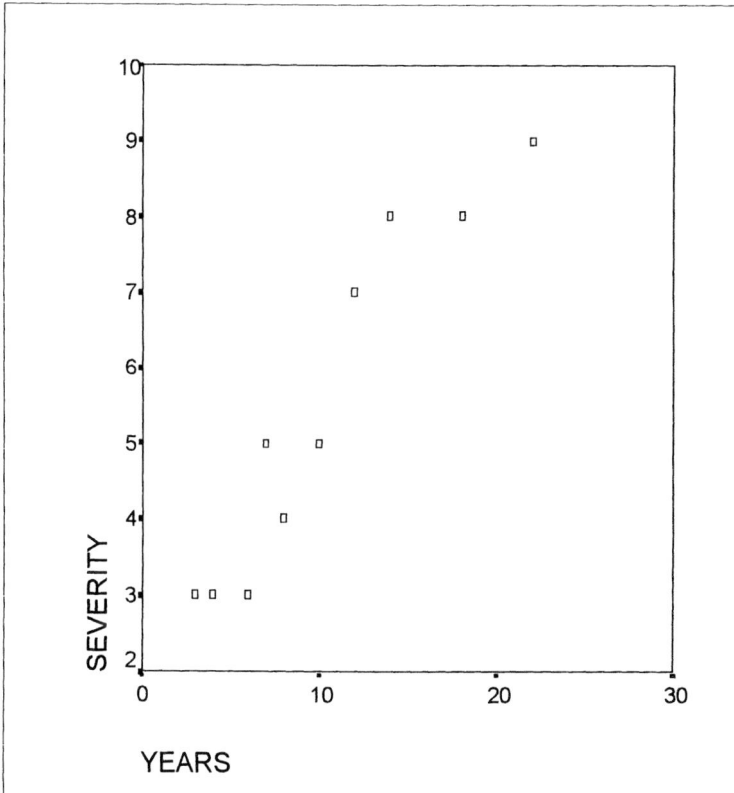

If someone had a crime severity of 6, for how many years would you predict the person would be sentenced to prison?

ADDITIONAL ACTIVITIES

1.   Rank the following correlations from lowest relationship to highest relationship.

.72   -.15   .52   .05   -.86   .55   -.44

2.   Rank the following correlations from lowest relationship to highest relationship.

.35   -.11   .52   .05   -.86   .75   -.44

3.   Interpret the following statements, in terms of both the strength and direction of the relationship

a. $r = -.35$ between GPA and family size
b. $r = .60$ between the number of students in a school and the school's budget
c. $r = .40$ between the number of hours of television viewing and a person's score on a test of current events knowledge.

4.   A researcher was interested in the relationship between the hours of sleep someone gets each night and their performance on a test of alertness. The following summary results were obtained:

	Hours of Sleep (X)	Score on Alertness Test (Y)
M	6.7	78.0
Sum	670	780
SS	26.1	1310
S^2	2.9	145.56
S	1.70	12.06

N=10

$\Sigma(X-M_X)(Y-M_Y)=119$

a.   Calculate the covariance.
b.   Calculate the correlation.

c.   Calculate the prediction equation using score on the alertness test as the dependent variable.

d.   Based on the prediction equation you calculated in part c, and with "SCORE" being the dependent variable, substitute the values of 8 and 10 for the amount of hours spent sleeping.

e.   Based on your answers to part d, draw your regression line on the following scatterplot.

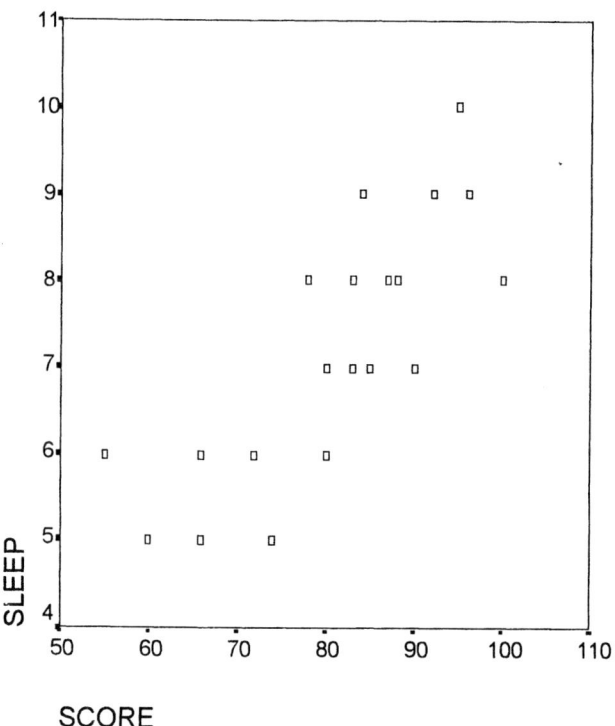

MODULE VII: SAMPLES, SAMPLING DISTRIBUTIONS, CONFIDENCE INTERVALS, AND EFFECT SIZES FOR ONE SAMPLE TESTS WITH N>100

SAMPLING AND ESTIMATION

In Module I we noted that the goal of inferential statistics is to make decisions about a population based on information that we gather from a sample. So we first obtain what is called a random sample. A truly random sample implies that all members or events of a particular population had an equal chance of being chosen for our sample. How do we do this? Imagine that we were interested in taking a random sample of 100 students from a university in order to gain information about their political affiliation. A true random sample would involve getting a list of all students at the university, and then randomly picking 100 of them. Perhaps this could be done by having all of their names on sheets of paper, putting these names in a hat, and then pulling out 100 of them.

Behavioral scientists are, unfortunately, very lax in the use of true random sampling. By that, I mean that it is rarely done. For example, at my university we have what is called an Introductory Psychology Research Requirement, which means that all undergraduate students taking

Introduction to Psychology are required to take part in a few research experiments. They do this by signing up for a project that interests them. Therefore, the researchers running these projects are not taking a random sample, but rather obtaining a sample of convenience. This is how the majority of research studies at universities, and even designated research organizations, are conducted. While such a technique does make life a great deal easier for the researcher, it brings with it the problem of our sample not being truly representative of the population of interest. Ssocial and behavioral scientists hope to make inferences about the population from a particular sample. If that sample consists only of undergraduate college students, taking Introduction to Psychology, at a particular university, who decide to participate in a research study based on how interesting it sounds, then how representative do you think that sample is of the true population? In actuality, not very.

SAMPLING DISTRIBUTIONS

Let's assume we now have our sample (random or otherwise). One of the most common statistical procedures is to find the mean of our sample on some variable of interest, and then make inferences concerning an unknown population mean. We call this unknown population mean μ (pronounced "mu").

It is important to note that with each new sample, we will get a new sample mean. That is to say, the mean that we get with one sample will most likely not be the same mean that we will get with another sample. So, there is some inherent error involved with any sampling situation, with the amount of error varying in terms of how many observations we make or sample size (N).

THE SAMPLING DISTRIBUTION OF THE MEAN

If we were to take a number of true random samples (note "true" random samples, meaning we obtained our random sample in the proper manner), each member or event within the sample has the same probability of being selected. Thus, we can associate a known probability with every possible sample and every possible value of the sample mean. That allows us to identify the probability distribution for the sample mean. Such a probability distribution is known as the Sampling Distribution of the Mean.

Let's look at an example to illustrate this point. Perhaps a university is concerned with keeping the number of students in each class low, so that there will be closer ties between the professor and the students. Let's focus for a moment only on Introductory Psychology classes, of which in a typical semester there may be 5 sections. Further, let's label these five sections as A, B, C, D, and E. These five sections will constitute the population, so N=5. If we wanted to take a random sample of two of these sections, there are ten different possible combinations:

Sections A-B	Sections A-C	Sections A-D	Sections A-E
Sections B-C	Sections B-D	Sections B-E	Sections C-D
Sections C-E	Sections D-E		

In a true random sample, each possible combination has the same probability of being selected, which in this case is 10%. Let's further assume that the number of students in each class on a particular day came out to be the following:

Section A	24
Section B	25
Section C	18
Section D	28
Section E	30

Therefore, there was a total of 125 students.

We have already learned how to compute the population mean and standard deviation, so we get the following figures:

$$\mu = \frac{\Sigma X}{N} = \frac{125}{5} = 25$$

$$\sigma = \sqrt{\frac{\Sigma(X-\mu)^2}{N}} = \sqrt{\frac{84}{5}} = \sqrt{16.8} = 4.1$$

Now we will take a look at the possible sample means that we would receive based on the samples that we draw.

Sections in sample	Probability of Sample	Sample Mean
A and B	10%	(24 + 25)/2 = 24.5
A and C	10%	(24 + 18)/2 = 21.0
A and D	10%	(24 + 28)/2 = 26.0
A and E	10%	(24 + 30)/2 = 27.0
B and C	10%	(25 + 18)/2 = 21.5
B and D	10%	(25 + 28)/2 = 26.5
B and E	10%	(25 + 30)/2 = 27.5
C and D	10%	(18 + 28)/2 = 23.0
C and E	10%	(18 + 30)/2 = 24.0
D and E	10%	(28 + 30)/2 = 29.0

Figure 7.1 represents the probabilities associated with obtaining a particular range of the sample mean, otherwise known as the sampling distribution of the mean.

Figure 7.1: Sampling Distribution of the Mean for Different Samples of Introductory Psychology Sections.

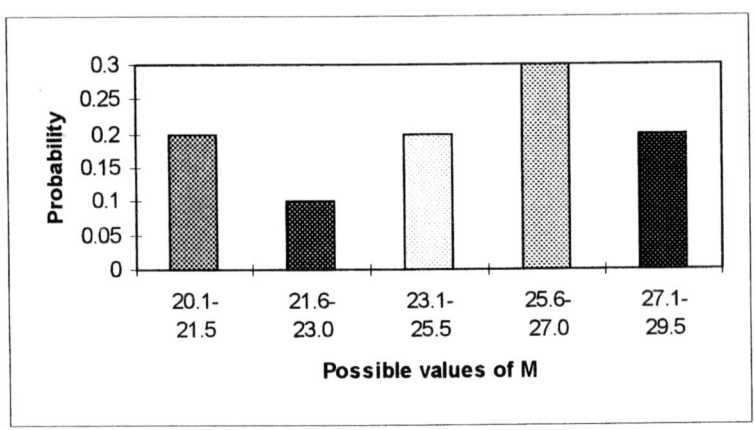

This is a relatively easy example, containing only 10 possible samples within the population. However, as more samples are taken, we will find that the distribution of the sample mean takes on more of a normal curve appearance.

There is a figure for a larger distribution in Figure 7.2.

<u>Figure 7.2</u>: Sampling Distribution of the Mean for 10,000 Means from a Normal Population with N=10 and N=50.

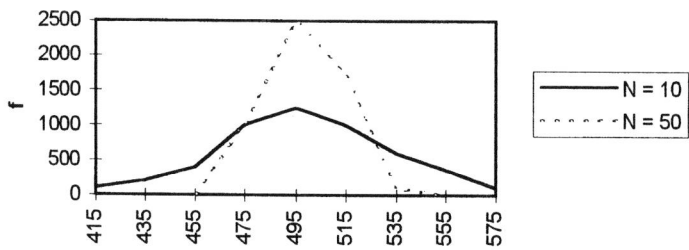

Here we can see that as more samples are taken (in this case, from N=10 samples to N=50 samples), that the normal distribution begins to graphically present itself.

We give the standard deviation of the sampling distribution of the mean a special name: The Standard Error of the Mean. It can be calculated as follows:

$$\sigma_M = \frac{\sigma}{\sqrt{N}}$$

What happens to the standard error of the mean as sample size increases? Let's assume that we have a distribution with a standard deviation of 10. Note in Table 7.1 how the standard error of the mean decreases as we increase the number of samples.

So we know that different samples will produce different estimates of a population, since these samples will consist of different individual means. Up until this point, we have been able to make direct comparisons between sample means and population means, because we have assumed that we know what the true population mean is. This is rarely the case. When the population mean (and associated distribution) is unknown, we rely on one of the most famous theorems in statistics: The Central Limit Theorem.

Table 7.1: The Relationship Between Sample Size (N) and Standard Deviation (σ) on the Standard Error of the Mean (σ_M).

$\underline{N}$	$\underline{\sigma}$	$\sigma / \sqrt{N}$	$\underline{\sigma}_M$
1	10	10 / 1.00	10.00
5	10	10 / 2.24	4.46
10	10	10 / 3.16	3.16
25	10	10 / 5.00	2.00
50	10	10 / 7.07	1.41
100	10	10 / 10.00	1.00
200	10	10 / 14.14	0.71
400	10	10 / 20.00	0.50
1000	10	10 / 31.62	0.32

THE CENTRAL LIMIT THEOREM

The central limit theorem can be stated as follows: In selecting random samples of size N from a population with mean μ and standard deviation σ, the sampling distribution of M approaches a normal probability distribution with mean μ and standard deviation $\sigma/\sqrt{N}$ as the sample size becomes larger, regardless of the shape of the parent distribution. What this basically tells us is that the more samples we take, the better our sample distribution will more closely approximate the true population distribution. Furthermore, the sampling distribution becomes nearly normal in shape even with relatively small sample sizes.

INFERENTIAL STATISTICS AND HYPOTHESIS TESTING WITH N>100

Suppose we know that the mean high school GPA for incoming freshman college students is 3.0. Now suppose that someone in the Admissions Office of University X wants to know if the mean high school GPA for students entering University X is significantly different from 3.0. From student records, it has been determined that this GPA is 3.1, acquired from a sample of 200 records.

The hypothesis we wish to evaluate is that the mean of the

population of students entering University X is the same as the mean of all students. We call this our null hypothesis, and express it as:

H_o: $M = \mu$

The null hypothesis is a way of stating that there will be no differences. If it turns out that there truly are no differences, we say that we fail to reject the null hypothesis. If it turns out that there are differences, we say that we reject the null hypothesis in favor of an alternative hypothesis.

If we decide that it is unlikely that the null hypothesis is true, we will reject the null hypothesis in favor of an alternative hypothesis, which, in this example, could take any of three forms:

H_1: $M < \mu$
H_1: $M > \mu$
H_1: $M \neq \mu$

The first two alternative hypotheses are called directional alternative hypotheses, while the last alternative hypothesis is called a non-directional alternative hypothesis. Directional alternative hypotheses are appropriate when the only type of outcome you care to detect is in one direction or the other of the null hypothesis. We choose to test the null hypothesis against a non-directional alternative hypothesis in this example because we would be interested in knowing whether the GPA of University X students is either higher or lower than the general population. At this point, you may be saying to yourself "obviously it is higher, because 3.1 is greater than 3.0." However, you must remember that we make the decision to use either a directional or non-directional hypothesis before we know any of the means. This is an important theoretical issue that we will touch on later, but will be covered more completely in a course on experimental research design.

To perform our test of the null hypothesis, we need the sampling distribution of the mean for $N = 200$, $\sigma = 0.5$. For hypothesis testing, we begin by assuming that the null hypothesis (there are no differences) is true. We then seek to determine how likely our sample result would be under that assumption. If the sample result would be extremely unlikely assuming that the null hypothesis is true, we reject null hypothesis in favor of the alternative hypothesis.

Assuming that the null hypothesis is true, the sampling distribution of the mean will have a mean of 3.0. We do not yet know the population standard deviation for this group, but with a sample size of 200 we may safely assume that our sample estimate of the population variance is, for all practical purposes, the same as the population variance. Even if GPA scores were not distributed approximately normally, the central limit theorem would still allow us to be confident that the sampling distribution of the mean will be normal, given our sample size of 200.

The next step in the process of testing our null hypothesis is to decide what will constitute a statistically significant difference between the two means. Just how unlikely does the result have to be before we reject the null hypothesis in favor of the alternative hypothesis? While we will discuss this point in more in future modules, scientific convention is to set our significance level at 0.05, meaning that if the probability of the sample outcome, under the assumption that the null hypothesis is true, is less than 1 in 20, we will reject the null hypothesis in favor of the alternative hypothesis. This 0.05 level is called the alpha level, as is designated as follows: $\alpha=.05$.

We are now ready to conduct our test. It is always a good idea to list what information has been given to us in the problem. Thus far, we know:

$M = 3.1$
$\mu = 3.0$
$\sigma = 0.5$
$N = 200$
$\alpha = 0.05$

To calculate the standard error of the mean, we enter the information into the formula as follows:

$$\sigma_M = \frac{\sigma}{\sqrt{N}} = \frac{0.5}{\sqrt{200}} = \frac{0.5}{14.14} = 0.04$$

To determine the probability of this outcome, assuming that the null hypothesis is true, we must standardize our scores. To do this we calculate a z score. Notice that this is a different formula for the z score than the one presented in Module V. However, the logic is the same. We are taking a score of interest, subtracting that score from an already established mean, and then dividing by a standardized value.

$$z = \frac{M - \mu}{\sigma_M}$$

$$= \frac{3.1 - 3.0}{0.04}$$

$$= \frac{0.1}{0.04}$$

$$= 2.50$$

We now need to determine the probability (or likelihood) of obtaining a z score of 2.50. By returning to the z-score table, we can see that p ($|z| \geq 2.50$) = .0062.

The probability level of the test (.0062) is less than our significance level (α) of 0.05, so we reject the null hypothesis, accept the alternative hypothesis, and conclude that the observed sample mean of 3.1 is statistically significantly greater than 3.0. Keep in mind that the probability value that we receive from the z-score table is the probability that the null hypothesis is true.

At this point, there are two things you should note. First, we have taken the absolute value of z. It is possible for a z score to be negative (which would have happened in the above example if the means were inverted, such that we were subtracting 3.1 from 3.0). Since our normal curve is symmetric, we need only deal with one side of the curve, so we choose to take the absolute value.

Second, are we 100% certain that our decision to reject the null hypothesis is correct? No, there is a chance (0.0062) that we could have observed a difference between these two means, even if the null hypothesis were true. However, the chances of this happening are so small (0.62%) that we will reject the null hypothesis in favor of the alternative hypothesis.

ACTIVITY 7.1: Suppose that the mean on an I.Q. test is known to be 100 for the general population. We are interested in whether people who have been given a new intelligence enhancing drug score significantly higher or lower than the typical mean I.Q. score. We obtain a random sample of 150 individuals, administer an I.Q. test to them, and find that the mean is 105 with a standard deviation of 15. The summary statistics are below:

M = 105
μ = 100
σ = 15
N = 150
α = 0.05

What is the null hypothesis and what is the alternative hypothesis for this problem?

H_0:

H_1:

The next step is to find the standard error of the mean, knowing that

$$\sigma_M = \frac{\sigma}{\sqrt{N}}$$

We must now standardize the scores into a z-score format, knowing that

$$z = \frac{M - \mu}{\sigma_M}$$

Given the z-score, what is the probability that our sample of individuals who took the intelligence enhancing drug does not have a significantly different I.Q. score than the typical population?

Do we accept the null hypothesis or reject it? Why?

ACTIVITY 7.2: Based on the information given in Activity 7.1, what would be the outcome if we found that the mean I.Q. score for our intelligence enhancing drug group was 101?

CONFIDENCE INTERVALS

At some point you may have been watching television or reading the newspaper and came across a statement such as this: "Eighty-five percent of the American public, plus or minus 3 percent, is in favor of governmentally funded health care." Statements such as this are based on the properties of a confidence interval. A confidence interval will give us a range of scores, within which we will be confident (to a particular confidence interval level) the true value of some dimension lies. In the above example, for instance, we might say that we are 95% confident that the true range of the American public in favor of governmentally funded health care is from 82% to 88%.

The use of a confidence interval allows us to more clearly characterize the state of our data. From Module III, you may recall that we do not use only one measure of central tendency to describe our data, we use three. The same is true for hypothesis testing. The more information you can calculate (such as a confidence interval, or an effect size which we will learn shortly), the better off you are.

Let's return to our example concerning the difference in GPA between the general population and a sample of students at University X. Let's also refresh ourselves with the 68-95-99.7 Percent Rule. In a standardized distribution, 68% of the scores will range between plus and minus one standard deviation, 95% of the scores will range between plus and minus two standard deviations, and 99.7% of the scores will range between plus and minus three standard deviations. The value associated with the standard error of the mean can take the place of standard deviations in problems such as this. Therefore, the 68% confidence interval will be the mean plus or minus one standard error of the mean, the 95% confidence interval will be the mean plus or minus two standard errors of the mean, and so on.

In our example, we calculated the standard error of the mean, and found it to be 0.04. The mean was 3.1. So, our confidence intervals (CI) would be as follows:

$$68\% \ CI = 3.1 \pm 0.04$$
$$95\% \ CI = 3.1 \pm 0.08$$
$$99.7\% \ CI = 3.1 \pm 0.12$$

Notice that the more certain we wish to be, the wider the confidence interval.

The three confidence intervals we have chosen to calculate have been simplified by the fact that they correspond to whole standard errors of the mean. To calculate any confidence interval, the following formula is used:

$$CI = M \pm (z\frac{1}{2}_{\alpha})(\sigma_M)$$

The z-score refers to the critical value cutting off the extreme α proportion of the sampling distribution.

Instead of estimating the 95 percent confidence interval from standard error of the mean, we can use the above formula. The outcome would be:

$$CI = M \pm (z\frac{1}{2}_{\alpha})(\sigma_M)$$

$$95\% \ CI = 3.1 \pm (1.96)(0.04)$$

$$95\% \ CI = 3.1 \pm .08$$

The 95% CI would range from 3.02 to 3.18.

ACTIVITY 7.3: If the 99 percent confidence interval were computed from the data in the above GPA example, what would it be?

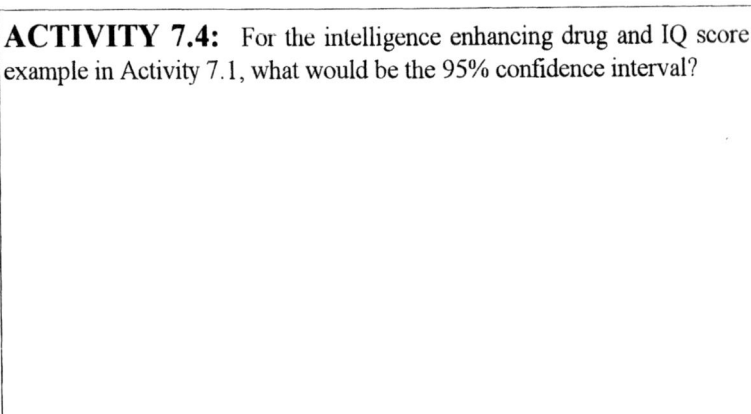

ACTIVITY 7.4: For the intelligence enhancing drug and IQ score example in Activity 7.1, what would be the 95% confidence interval?

EFFECT SIZE

The effect size is an important adjunct to other inferential procedures because it addresses the issue of how big, or important, the observed effect (the difference between your two means) is. Because the standard error will get smaller as the sample size gets larger, larger sample sizes result in decisions to reject the null hypothesis more frequently than smaller samples, other things being equal. Therefore, with a really large sample size, you might reject the null hypothesis when there is a very small difference between your two means of interest.

Such a situation brings about the distinction between statistical significance and meaningful significance. For example, you may find that on a particular biology exam that there is a statistically significant difference between the scores of males and females, with females having a mean score of 85 and males having a mean score of 84. This is only a one point difference; however, if your sample size is large enough and your standard error of the mean is small enough, this could potentially be a significant difference. It is then up to the researcher to decide whether this one point difference is a meaningful difference. By calculating an effect size, that decision is made much easier.

The effect size, denoted by the letter d, is calculated as follows:

$$d = \frac{M - \mu}{\sigma}$$

In our study of GPA scores for students at University X, the effect size measure (d) is:

$$d = \frac{M - \mu}{\sigma}$$

$$= \frac{3.1 - 3.0}{0.5}$$

$$= \frac{0.1}{0.5}$$

$$= 0.2$$

Effect sizes are characterized as follows:

  0.20 and below = small
  0.35 to 0.50 = moderate
  0.60 and above = large

Therefore, in addition to saying that students who attend University X have significantly higher GPAs than the average population, we can note that this is a small effect.

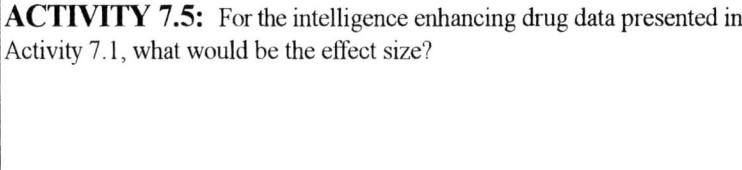

ACTIVITY 7.5: For the intelligence enhancing drug data presented in Activity 7.1, what would be the effect size?

ACTIVITY 7.6: For the intelligence enhancing drug data presented in Activity 7.2, what would be the effect size?

ADDITIONAL ACTIVITIES

1.   When entering the city of Wheeling, West Virginia you will come across a sign that reads: "Welcome to Wheeling -- The Friendly City." A social psychologist was interested in determining if Wheeling really was "The Friendly City." She obtained a sample of 300 Wheeling residents, and had them fill out a questionnaire concerning friendliness. In the general population, the mean friendliness score is 30 with a standard deviation of 10. The mean of the Wheeling sample was 32.

   a.   Find the probability of obtaining such a sample mean from the aforementioned population.
   b.   Calculate the 95% confidence interval.
   c.   Calculate the effect size.
   d.   Describe the results of this study, indicating whether or not this is a statistically significant result, what the confidence interval tells us, what the effect size tells us, and what we can conclude about the friendliness of Wheeling residents in relation to the general population.

2.   A social worker decided that his clients lacked an important skill - assertiveness. He spent the next eight weeks teaching assertiveness during group sessions. A psychologist friend of his suggested that he determine whether the training was successful by testing his group with a popular Assertiveness Test, the population mean of which is 30 with a standard deviation of 12. The 150 clients took the test, and the statistics that follow were obtained. with a mean of 30.5

   a.   Find the probability of obtaining such a statistic from the aforementioned population.
   b.   Calculate the 95% confidence interval.
   c.   Calculate the effect size.
   d.   Describe the results of this study in terms understandable to the social worker's supervisor, indicating whether or not this is a statistically significant result, what the confidence interval tells us, what the effect size tells us, and what we can conclude about the assertiveness of the clients.

MODULE VIII: PRINCIPLES OF STATISTICAL POWER ANALYSIS AND DECISION MAKING

Previously, we noted that it is a statistical convention to place the level of significance for rejection of the null hypothesis at $\alpha=0.05$. We will now take a closer look at what this value really does for us in relation to our null and alternative hypotheses.

TYPES OF DECISION ERRORS

As researchers, there are two types of errors that we can make when we perform our statistical tests, Type I errors and Type II errors.

TYPE I ERRORS

If it is the case that the null hypothesis is true (i.e., there are no significant differences), there is an α chance that you will incorrectly reject the null hypothesis. This is called a Type I error, although some statisticians may refer to it as an alpha (α) error. If, because of convention, we set $\alpha = .05$, then the probability of a Type I error is .05. Since all probabilities must be equal to 1 (or 100%), there is a $1 - \alpha$ chance that you will correctly accept, or fail to reject, the null hypothesis. If $\alpha = .05$, then $1 - \alpha$ is .95.

TYPE II ERRORS

If it is the case that the alternative hypothesis is true (i.e., there is a significant difference), there is some definable probability that you will fail to reject the null hypothesis, thus making an incorrect decision. This is called a Type II error, although some statisticians may refer to it as a beta (β) error. Again, since all probabilities must be equal to 1 (or 100%), there is a $1 - \beta$ chance that you will correctly reject the null hypothesis. For the most part we are unable to set β. However, its role in statistical decision making is quite important, since the value represented by $1 - \beta$ is equal to the power of the statistical test. Thus, the power of the statistical test ($1 - \beta$) is the probability of successfully detecting a significant difference when one actually exists.

The relationship between the decision error and their probabilities are shown in Table 8.1.

Table 8.1:   Relationship Between Decision Error and Their Probabilities.

	IN REALITY...	
	H_0 IS TRUE	**H_1 IS TRUE**
YOU DECIDE TO...		
ACCEPT H_0	Correct Decision $1 - \alpha$	Type II Error β
REJECT H_0	Type I Error α	Correct Decision $1 - \beta$

PRINCIPLES OF STATISTICAL POWER ANALYSIS

There are certain interactions and trade-offs of which you should be aware when determining the power of your statistical test.

● The probability of making a Type II error increases as α decreases. Likewise, as the probability of making a Type II error increases, power $(1-\beta)$ decreases. This is the main reason for picking an intermediate value for α (such as 0.05). The lower we set α, the lower our power. However, this is a two-way street--despite having decreased power, we also decrease the probability of making a Type I Error.

● The probability of making a Type II Error decreases as sample size (N) increases. Therefore, the larger your sample size, the more powerful the statistical test. A larger sample size will also capitalize on the central limit theorem, thus making your sample more representative of the true population.

● The probability of making a Type II Error increases as the variance of your sample increases. Therefore, it is to your advantage to control or account for as much variance as possible. If you can decrease the "noise" component of an effect, you will have a more powerful test.

● The probability of making a Type II Error decreases as the effect size (d) increases. There is little that we can do to increase the effect size, but decreasing the "noise" will have some impact.

MODULE IX: SAMPLES, SAMPLING DISTRIBUTIONS, CONFIDENCE INTERVALS, AND EFFECT SIZES FOR ONE SAMPLE TESTS WITH N<100

The z statistic that we covered in Module VIII is fine to use when we have a large sample size. That is because when we have a large sample size, the standard error of the mean distributes itself as a normal distribution, and we are able to capitalize on such a distribution, primarily through being able to utilize the 68-95-99.7 rule. While there is some contention among statisticians as to what actually constitutes a large sample size, we will assume that any data set with more than 100 observations can be considered large.

THE t DISTRIBUTION

The standard error of the mean does not distribute itself normally when our sample size is small (e.g., N < 100). Therefore we need a different sampling distribution and test in those situations. This new sampling distribution is called the t distribution. When we actually perform the statistical test, we are performing what is called a t-test. A t-test is calculated as follows:

$$t = \frac{M - \mu}{\sigma'_M}$$

with $\sigma'_M = \sqrt{\dfrac{\sigma^2}{N}}$

The t distribution has a denominator that is influenced by sampling error. Also note that we are using the character σ'_M to indicate that this is an estimate of the true standard error. This is necessary due to having a smaller sample size, which leads to not being as certain about what the true standard error is.

The shape of the t distribution varies as a function of a quantity known as the degrees of freedom (df). The t distribution is similar to the standard normal distribution, but there is a different distribution (and thus a different shape of the curve) for every different value of the degrees of freedom (df). The degrees of freedom are calculated as N-1.

So what is this new term, known as the degrees of freedom? The "freedom" in degrees of freedom refers to the freedom of any number to have any possible value. If you were asked to pick two numbers and there were no restrictions, both numbers would be free to vary (take on any value) and you would have two degrees of freedom. What if a restriction was imposed, for example, that $\Sigma X = 0$. Then one degree of freedom is lost because of that restriction; that is, when you now pick the two numbers, only one of them is free to vary so that you are able to meet the restriction.

Imagine the following example to illustrate this point: You have the following three values in a data set: {1, 2, 3}. In addition, there is a restriction imposed, such that $\Sigma X = 6$. Any two of your three data set values can vary randomly, but you must maintain control over one of these data set values in order to meet the imposed restriction. Therefore, if your score of 1 were to vary and become a 4, and your score of 2 were to vary and become a 0, you could alter the value of your score of 3 to a 2, such that you can meet the restriction of $\Sigma X = 6$.

In the statistical tests that we will now be considering, there is a restriction built in such that $\Sigma(X - M) = 0$, and in order to meet that requirement, one of the X's is determined. All X's are free to vary except one, and thus we will typically denote degrees of freedom as N-1 (or the total number of data values that we have, minus one).

There are three situations where we might use a t test. The first is called a one-sample test, which is what this module will cover. A one-sample test is used when you are comparing an obtained sample mean of N<100 to an already established population mean. The other two types of situations consist of comparing two sample means in various ways, and will be discussed in the following two modules.

TESTING HYPOTHESES

The proceedures that we will be using for the t-test are almost identical to those used with the z-test in Module VII. Let us start by actually walking through an example.

Suppose we know that the mean number of children in all families is 2.5, and we wish to make an inference about the size of Catholic families. We obtain a random sample of 45 Catholic families and observe the mean number of children is 3.11 and that the estimated population variance (σ^2) is 4.00. This is considered a one-sample test, because we have only collected data concerning one sample (in this case, Catholic families), and will be comparing that mean to an already established population mean (in this case, all families).

THE FIVE STEP PROCESS

There are five steps in making a statistical test of significance.

Step 1: Hypotheses. Here we present the null and alternative hypotheses for our statistical test. For the null hypothesis, we would expect that there would be no statistical difference between the number of children in Catholic families and the number of children in the general population. As a non-directional alternative hypothesis, we would expect that there is a difference between these two means. Our null hypothesis and alternative hypothesis would be expressed as follows:

H_0: $M = \mu$
H_1: $M \neq \mu$

Step 2: Sampling Distribution. Here we have fewer than 100 observations in our study, so we will not be using the z distribution, which is appropriate only if we have a large sample size. Instead we will be using the t distribution. Using the t distribution, we see that with 45 observations, our degrees of freedom will be N-1, or 45-1 = 44.

Step 3: Level of Significance. Here we decide at what level our test will be considered statistically significant. The t score corresponding to this level is known as the critical value of t, and is denoted as t'. As we previously mentioned, convention is to set α=.05. If we turn to a t distribution table (located in the back of your manual), look across the table for α=.05 for a two-tailed test, and down the table for 44 degrees of freedom, we see that there is no 44 degrees of freedom on our table. When you have a distribution table that does not include your actual degrees of freedom value, always go to the next lower degrees of freedom value. In our table, that would mean that we would go down to 40 degrees of freedom. So, with 40 degrees of freedom, our critical value is 2.021, which we designate as follows: t'(40) = 2.021. Notice the prime sign after the t, showing that it is a critical value, and that the number in parentheses indicates the degrees of freedom associated with this critical value.

There are two things of note at this point. First, why do we choose a two-tailed test, rather than a one-tailed test? A two-tailed test is used when we choose a non-directional hypothesis, like we have chosen for this problem. It implies that we have no specific pre-conceived notion about whether our sample mean will be greater than or less than the established population mean, only that we believe they might be different. In almost all cases, a two-tailed test procedure will be used. A one-tailed test is used when we choose a directional hypothesis, and have a pre-conceived notion about the particular way our scores will be different (e.g., the sample mean will be larger than the population mean, or that the sample mean will be smaller than the population mean). You may say, "But I can tell from these data that the sample mean is larger than the population mean, so why can't I use a one-tailed test?" The use of either a one- or two-tailed test is determined before you see any of the means (usually, before any data are collected), so it is a violation of the integrity of statistical testing to decide post-hoc (after the fact) to use one type of test over the other. Therefore, most researchers opt for the non-directional hypothesis and the two-tailed statistical test.

Second, what does this critical value (t') really mean? Soon we will perform the actual statistical test of our null hypothesis (called the t-test). If the value we receive from the statistical test is equal to or greater than our critical value, then we say that the test is statistically significant, and we will reject the null hypothesis in favor of the alternative hypothesis. If the value we receive from the statistical test is less than our critical value, then we say that the test is not statistically significant, and we must accept the null hypothesis. Therefore, the critical value gives us a point at which our actual statistical test will either be statistically significant or not.

Step 4: Perform the Statistical Test. We know that the t-test formula for this problem is:

$$t = \frac{M - \mu}{\sigma'_M}$$

First we calculate σ'_M, which in our example is:

$$\sigma'_M = \frac{\sigma^2}{N}$$

$$= \frac{4}{45}$$

$$= .089$$

$$= .298$$

Our t test is as follows:

$$t = \frac{3.11 - 2.5}{.298}$$

$$= 2.047$$

Step 5: Decision and Interpretation. Our t-test result was 2.047 and our critical value (t') was 2.021. Therefore, our actual t-test result was greater than the critical value, making this a statistically significant test, and we should reject the null hypothesis (that there is no statistically significant difference between the means) in favor of the alternative hypothesis (that there is a statistically significant difference between the means). What we can conclude from this study, then, is that there is a statistically significant difference in the mean of our study participants from the mean of the population. To indicate this outcome, we use the following format: t(44)=2.047, p<.05. This indicates that the t-test value, with 44 degrees of freedom, came out to be 2.047. By indicating p<.05, we are showing that the probability of the null hypothesis being true is quite small (less than 5%, which is what we set as our alpha level).

Finally, we must note the direction of this effect. By noting the direction of the effect, we indicate which mean was higher or lower. Therefore, we can say that the mean number of children in Catholic families (3.11) is significantly larger than the mean number of children in the general population (2.5).

ACTIVITY 9.1: Suppose we know that the mean number of hours of television viewed by Americans is 25, and we wish to make an inference about the number of hours WJU students watch television. We obtain a random sample of 35 students and observe the mean number of television hours viewed is 28, and that the estimated population variance is 9.6. Perform the five steps for analysis.

    1.   Hypotheses

    2.   Sampling Distribution

    3.   Level of Significance

4.   Perform the Statistical Test

5.   Decision and Interpretation

CONFIDENCE INTERVALS

The confidence interval is calculated very much the same as when we performed the z-score test confidence interval. However, we will be inserting the alpha value for t, rather than z. In general, the formula is

$$CI = M \pm t\frac{1}{2}_{\alpha}\sigma'_M = 1 - \alpha$$

For the 95% confidence interval in our example, we would calculate as follows...

$$CI = 3.11 \pm (2.021)(.298) = 1 - .05$$

$$CI = 3.11 \pm .602 = .95$$

$$CI = 2.508 \text{ to } 3.712 = .95$$

So, in this study we found M=2.5, which is not within our 95% confidence interval. That is one reason that we said M and μ are significantly different.

ACTIVITY 9.2:   For the data from Activity 9.1, calculate the 95% confidence interval.

EFFECT SIZE

The effect size is computed exactly as for the previous z score test of a single mean. Remember that $\sigma' = \sqrt{\sigma^{2'}}$. For the present problem,

$$d = \frac{M - \mu}{\sigma'}$$

$$d = \frac{3.11 - 2.5}{2}$$

$$d = 0.31$$

Thus, we can characterize this as a small to moderate effect size.

ACTIVITY 9.3: For the data from Activity 9.1, calculate the effect size.

ADDITIONAL ACTIVITIES

1.   Suppose that the mean on an I.Q. test is known to be 100 for the general population. We are interested in whether a particular group of children in a kindergarten class score significantly higher or lower than the typical mean I.Q. score. We obtain a sample of the 20 children, administer the I.Q. test to them, and get the following results: M=104 and σ=10.

   a.   What are the null and alternative hypotheses for this situation?
   b.   What is the estimated standard error of the mean?
   c.   Perform the appropriate statistical test, with α=.05.
   d.   Describe the results of your findings. Is this test statistically significant? Why or why not? What is your decision concerning the acceptance or rejection of H_0 and H_1?
   e.   Compute the 95% confidence interval.
   f.   Compute the effect size.

2.   Many manufacturers of light bulbs advertise their 75 watt bulbs as having a life of 750 hours (about a month). Suppose two electrically minded students wired up 12 bulbs so that the time was recorded when they burned out. The following statistics were obtained: M=745, σ=5

   a.   What are the null and alternative hypotheses for this situation?
   b.   What is the estimated standard error of the mean?
   c.   Perform the appropriate statistical test, with α=.05.
   d.   Describe the results of your findings. Is this test statistically significant? Why or why not? What is your decision concerning the acceptance or rejection of H_0 and H_1?
   e.   Compute the 95% confidence interval.
   f.   Compute the effect size.

MODULE X: INFERENCES ABOUT TWO INDEPENDENT SAMPLE MEANS

In Module IX, we examined a one-sample t-test, which is used when comparing one sample of data (with N<100) to an already established population. In this module, we will be discussing how to determine whether there is a statistically significant difference between two independent sample means. By independent, it is implied that we are comparing two separate groups that do not overlap in terms of the numbers which make up their scores. For example, you can't have blue eyes and brown eyes at the same time, so eye color would constitute independent grouping. You also can not be a freshman and a senior at the same time, so class standing would constitute independent grouping. Behavioral scientists also place people into two different situations in an attempt to determine the effects of each situation (the independent variable) on some sort of score (the dependent variable). In this case, the groups in which the people were placed would constitute independent groups. The end goal, then, is to determine whether the mean of Group 1 is significantly different from the mean of Group 2.

TESTING HYPOTHESES

The null hypothesis and alternative hypothesis for the test of an independent mean difference can be expressed in two ways. Both of the following options imply that there will not be a difference between the two group means (null hypothesis) and that there will be a difference between the two group means (non-directional alternative hypothesis).

Option 1: H_0: $M_1 = M_2$
              H_1: $M_1 \neq M_2$

Option 2: H_0: $M_1 - M_2 = 0$
              H_1: $M_1 - M_2 \neq 0$

If you were to use a directional alternative hypothesis, H_1 could be substituted with one of the following:

H_1: $M_1 > M_2$ or H_1: $M_1 < M_2$

In Module IX, our t-test formula was equal to the following:

$$t = \frac{M - \mu}{\sigma'_M}$$

Basically, we were taking the difference between a sample mean and a population mean, and dividing by some level of error measurement (the standard error of the mean). The same will hold true when we are comparing two independent sample means. However, the t-test formula takes on a new look, as follows:

$$t = \frac{M_1 - M_2}{\sigma'_{M1-M2}}$$

with

$$\sigma'_{M1-M2} = \sqrt{\frac{[(N_1-1)\sigma^2_1 + (N_2-1)\sigma^2_2]}{(N_1+N_2-2)} * \left(\frac{1}{N_1} + \frac{1}{N_2}\right)}$$

and degrees of freedom = $N_1 + N_2 - 2$, which is the same as (N_1-1) + (N_2-1).

Again, we are looking for the difference between two means (the mean of sample one versus the mean of sample two), and dividing by a measure of error known as the standard error of the mean difference.

Pay particular attention to the following sub-formula:

$$\frac{(N_1-1)\sigma^2_1 + (N_2-1)\sigma^2_2}{(N_1+N_2-2)}$$

This is known as the pooled variance. The pooled variance is a measure of the variance based on a weighting of how many observations are taken for each group, and is pooled (added) together. We do this because, should we have groups with different numbers of observations, we want to give more weight (or significance or credence) to the group that has more observations since the greater the number of observations, the more representative the sample. We will be using the pooled variance later in this module.

AN EXPERIMENTAL EXAMPLE

Let's try an example. To investigate the effects of room color on eating behavior, a researcher arranges for volunteers to report to the laboratory to participate in what they think will be a taste test for 10 newly invented food products. Forty students sign-up for the experiment and twenty are randomly assigned to one of two groups. Students in group one are allowed to eat as much of the foods as they want while seated in a black-painted room. Students in group two are also allowed to eat as much of the foods as they want, but they are seated in a yellow-painted room. After 5 minutes, the experimenter records the number of grams of food consumed under the two testing situations. Table 10.1 contains some summary information.

We will continue to use the 5 step process in order to determine statistical significance.

Step 1: Hypotheses. In this case, we are interested in seeing whether there is a significant difference in the amount of food eaten between the two testing situations (different colored rooms). We can state the null and alternative hypotheses in one of two ways

$$H_0: M_1 = M_2 \quad \text{and} \quad H_1: M_1 \neq M_2$$

or

$$H_0: M_1 - M_2 = 0 \quad \text{and} \quad H_1: M_1 - M_2 \neq 0$$

Step 2: Sampling Distribution. We have two independent groups, since you could not be in both Group 1 and Group 2 at the same time.

Table 10.1: Summary Data for Room Color and Food Consumption.

	Group 1 (Black Room)	Group 2 (Yellow Room)
Raw Scores	44.00	47.00
	49.00	52.00
	47.00	50.00
	45.00	48.00
	45.00	48.00
	49.00	52.00
	49.00	52.00
	48.00	50.00
	48.00	52.00
	50.00	53.00
	47.00	52.00
	46.00	53.00
	47.00	52.00
	47.00	54.00
	47.00	53.00
	48.00	50.00
	44.00	44.00
	41.00	47.00
	42.00	48.00
	36.00	43.00
Mean	46	50
Variance	11.1	9.6
N	20	20

Therefore, we will be using the t distribution, with degrees of freedom = N_1 + N_2 - 2. So,

df = 20 + 20 - 2 = 38

Step 3: Level of Significance. Staying with convention, we set α = .05. Referring to the t distribution table (in the back of your manual), with

$\alpha=.05$ for a two-tailed test, and df=38, we get the following critical value: t'(30) = 2.042. Note that since df=38 did not exist that we went to the next lower number. After performing the statistical test, if the actual t score is equal to or greater than 2.042, we will reject the null hypothesis and state that there is a significant difference between the two group means. If the actual t score is less than 2.042, we will accept the null hypothesis and state that there is no significant difference between the two group means.

Step 4: Perform the Statistical Test. Our statistical test has the following form:

$$t = \frac{M_1 - M_2}{\sigma'_{M1\text{-}M2}}$$

with

$$\sigma'_{M1\text{-}M2} = \sqrt{\frac{[(N_1-1)\sigma^2_1 + (N_2-1)\sigma^2_2)]}{(N_1+N_2-2)} * (\frac{1}{N_1} + \frac{1}{N_2})}$$

Therefore,

$$t = \frac{46 - 50}{\sqrt{\frac{[(20-1)(11.1) + (20-1)(9.6)]}{(20+20-2)} * (\frac{1}{20} + \frac{1}{20})}}$$

$$= \frac{46 - 50}{\sqrt{\frac{[(19)(11.1) + (19)(9.6)]}{(38)} * (.05+.05)}}$$

$$= \frac{-4}{\sqrt{\frac{[(19)(11.1) + (19)(9.6)]}{(38)} * (.05+.05)}}$$

$$= \frac{-4}{\sqrt{\frac{(210.9 + 182.4) * (.1)}{38}}}$$

$$= \frac{-4}{\sqrt{1.017}} = \frac{-4}{1.01} = -3.96$$

Step 5: Decision and Interpretation. Focusing on the absolute value of our t-test result, we get t=3.96. Why should we focus on the absolute value? The reason we obtained a negative score was because we put the smaller group mean into our equation first. Had we put the smaller group mean in second, we would have obtained a positive t value. Therefore, whether your t-value is positive or negative is an artifact of the order in which you place the means into the formula, and really plays no role in the decision to either accept or reject the null hypothesis.

Because our actual t-test score of 3.96 is greater than the critical value of 2.042, we therefore reject the null hypothesis and say that there is a statistically significant difference between the means of the two groups. More specifically concerning the direction of the effect, the group in the yellow room (M=50 grams) ate more food than the group in the black room (M=46 grams). The proper final form for this outcome would be: t(38)=3.96, p<.05.

ACTIVITY 10.1: Chem-Co Incorporated is testing a new drug which they believe will reduce the symptoms associated with Alzheimer's Disease, which is a debilitating degenerative disease of age characterized by decreased memory and cognitive ability. One group of 12 patients is given a placebo while another group of 12 patients is given the new drug. Following a drug therapy course of eight weeks, all patients are then tested. A higher score indicates a higher incidence of Alzheimer-like symptoms. Some summary information is below:

	Placebo Group	Drug Group
Raw Scores	42	36
	40	37
	38	29
	35	29
	41	40
	39	27
	45	40
	39	26
	43	40
	46	42
	40	36
	42	36
Mean	40.83	34.83
Variance	9.24	31.66
N	12	12

Perform the five steps to determine whether there is a statistically significant difference between the two groups.

1. Hypotheses

2. Sampling Distribution

3.   Level of Significance

4.   Perform the Statistical Test

5.   Decision and Interpretation

CONFIDENCE INTERVALS

The confidence interval for two independent groups is calculated as follows:

$$CI = (M_1 - M_2) \pm (t_\alpha)(\sigma'_{M1-M2}) = 1 - \alpha$$

For the 95% confidence interval,

$$CI = (46-50) \pm (2.042)(1.017) = 1 - .05$$

$$CI = -4 \pm 2.077 = 0.95$$

$$CI = -6.077 \text{ to } -1.923$$

Thus, in this experiment we found a 4 gram difference between the means of Group 1 and Group 2. If we were to perform this experiment over and over again, we could be 95% certain that true difference between the means will be somewhere between 1.923 grams and 6.077 grams.

ACTIVITY 10.2: Based on the information presented in Activity 10.1, calculate the 95% confidence interval.

EFFECT SIZE

Note the difference in the effect size formula from those previously encountered. In this case, we are using the square root of the pooled variance in the denominator.

$$d = \frac{M_1 - M_2}{\sqrt{\sigma'_{pooled}}}$$

In our example, effect size is calculated as follows:

$$d = \frac{M_1 - M_2}{\sqrt{\dfrac{(N_1-1)\sigma^2_1 + (N_2-1)\sigma^2_2}{(N_1+N_2-2)}}}$$

$$= \frac{46 - 50}{\sqrt{\dfrac{(19)(11.1) + (19)(9.6)}{38}}}$$

$$= \frac{-4}{\sqrt{10.35}} = \frac{-4}{3.22} = -1.24$$

Again, we are interested only in the absolute value of the test result, so we can say that the effect size is 1.24, and can characterize this as very strong.

ACTIVITY 10.3: Based on the information presented in Activity 10.1, calculate the effect size.

ADDITIONAL ACTIVITIES

1. An experimenter randomly divided volunteers into two groups. One group fasted for 24 hours and the other group fasted for 48 hours. Scores below represent the number of ounces of ice cream consumed during the first 15 minutes after the fast was over.

	24 Hour Fast Group	48 Hour Fast Group
M	15	19
σ^2	4.3	5.1
N	20	18

    a. Calculate the degrees of freedom.
    b. Calculate the standard error of the mean difference.
    c. What is t' with $\alpha=.05$?
    d. Perform the appropriate test to determine whether or not there is a statistically significant difference.
    e. Write a descriptive summary of your results, indicating whether or not this is a statistically significant test, and what conclusions can be drawn.

2. A certain psychology professor was interested in whether the presence of an attractive versus an unattractive research assistant had any influence on how many errors someone made on a test of word recall. Two groups of participants were given 2 minutes to memorize a list of 20 words and then were asked to recall as many of the words as they could. Group 1 participants were given the words to memorize by a very attractive research assistant, whereas Group 2 participants were given the words to memorize by a very ugly research assistant. The following data were compiled, with the mean indicating the number of words correctly remembered out of 30:

	Group 1 (Attractive)	Group 2 (Ugly)
M	23	20
σ^2	4.3	5.1
N	15	15

a.   Calculate the degrees of freedom.
b.   Calculate the standard error of the mean difference.
c.   What is t' with $\alpha=.05$?
d.   Perform the appropriate test to determine whether or not there is a statistically significant difference.
e.   Write a descriptive summary of your results, indicating whether or not this is a statistically significant test, and what conclusions can be drawn.

MODULE XI: INFERENCES ABOUT TWO RELATED SAMPLE MEANS

In contrast to independent sample means (and the use of an independent t-test), related sample means imply that the scores that we receive are in some way related. Some researchers may also refer to such tests as within group tests, correlated groups tests, repeated measures tests, or pre-post tests.

WHEN TO USE A RELATED MEANS DESIGN

There are three main ways that we could collect data for a related sample mean test.

1. <u>Pre-post designs</u>. In pre-post designs, we gather data on some aspect that interests us (pre-test), then we do something which we believe may alter these data, and then re-test the same group of people to see if their scores have changed (post-test). One typical example involves training programs. Perhaps we give a group of employees a test to determine how customer focused they are, and decide that based on their scores that they are low in customer focus. We may then implement a training program which addresses issues of customer focus. To determine whether or not our

program has worked, we then administer the same test to the same people and see if their scores have increased (or decreased). If their scores increase significantly, we may conclude that our customer focus program is effective.

2. Matched-pair designs. In matched pair designs, we match our observations based on some important criteria, such as age, gender, education, I.Q., weight, etc. We will still have two groups, but if age is important to us, we will make sure that the score of a 16 year old person in group one is compared to the score of a 16 year old person in group two.

3. Yolked designs. Yolked designs are less common, but you still may come across them in the behavioral sciences. In yolked designs, two different groups experience the same experimental situation, however, only one group has control over the experience. In addition, the group that has control over their experiences also has control over the experiences of the other group. For example, imagine that we have two groups of people in two identical rooms. Then we raise the temperature an equal amount in both rooms so that the temperature becomes uncomfortable. However, only one group has control over the thermostat to lower the temperature, and when they lower the temperature, they lower the temperature not only in their room, but also the room containing group two. Group two, in contrast, has no control over any aspect of the temperature.

ADVANTAGES AND DISADVANTAGES OF A RELATED MEANS DESIGN

Many times, a research study design could be conducted as either a independent sample design or a related sample design. For example, refer back to our example from the independent sample design from the previous module. In this example, we had our groups of participants sampling foods either in a black room or a yellow room. In theory, it would have been possible to have the same group of participants sample foods in both rooms.

There are, however, some advantages and disadvantages of choosing to use a related means design over an independent groups design.

ADVANTAGES

- Decreased sample size, Using the same participants in

both conditions is an advantage to a related means design, because you will need fewer participants in your study to obtain the same degree of statistical power. Why? Remember that power increases as variance decreases; therefore, the more observations we take from the same people, the more we are able to control for variance due to the way each individual person varies. We are not able to do that in an independent means study, since we only measure each person once.

● Change within the same group of people. Another advantage of a related means design is that we can see if there has been a change in the scores for each individual person in our experiment on a 1 to 1 basis, rather than inferring change from different groups of people.

DISADVANTAGES

● Practice effects. When we do something over and over again, we typically get better at it. Therefore, we may see some improvement in peoples' scores based merely on the fact that they are getting better on the task we are asking them to perform. If we get a statistically significant difference, we must then decide if this change in scores is because there is a true difference between our two conditions, or whether our participants are just getting better at the task.

● Testing effects. Similar to practice effects, testing effects are when our participants get familiar and comfortable with the procedures of the testing situation. Initially, our participants may be a little "wary" of any experimental situation; but after repeated exposures to the procedures, they realize that nothing bad is going to happen to them. This decrease in their uncertainty of the experimental situation (and possibly stress and anxiety) may influence subsequent measurements taken from them.

● Carryover effects. Often times something external to the research environment may cause a change in scores.

Imagine that I asked you to run around an athletic track as fast as you could, both on Monday and on Friday. On Monday I asked you to perform the run in shorts and on Friday I asked you to perform the run in sweatpants. Let's assume that you ran slower on Friday. Was it due to the change in apparel slowing you down? Perhaps, although you might also have fallen down a flight of stairs on Wednesday, and are still a little sore. This injury on Wednesday may also be influencing your Friday scores. Whenever you compare the scores of the same people on two different days, you need to take into consideration the possibility that other extraneous variables are playing a role in the outcome.

● Fatigue. Finally, the more times you have the same people do the same thing, the more tired and bored they become. Thus, their scores may be altered due to boredom or fatigue, rather than the experimental conditions to which your participants are exposed.

TESTING HYPOTHESES

The t-test for two related means can be calculated in various ways. One of the easiest is to perform the t-test at though it were a one sample test on difference scores. What makes this unique is that our dependent variable will be the difference between two paired observations.

We can write the null and alternative non-directional hypotheses for this test as follows:

H_0: Mean difference between the paired observations = 0
H_1: Mean difference between the paired observations ≠ 0

The formula for the test is as follows:

$$t = \frac{M_{DIFF}}{\sigma_{MDIFF}}$$

$$\text{with } \sigma_{MDIFF} = \sqrt{\frac{\sigma^2}{N}}$$

and degrees of freedom = N-1

In relation to the degrees of freedom, note that N represents the number of pairs of observations that we have, and not the total number of observations.

AN EXPERIMENTAL EXAMPLE

Suppose we are interested in whether or not a course on religion has any impact on how religious people are. We might assume that by subjecting students to a religion course, that they will become more religiously-minded in the future. So, at the beginning of the semester we assess how religious the students are by asking them to complete a questionnaire designed to measure religious values. Then, during the last week of the religion course, we have them complete the same questionnaire again to see if there are any changes in their religious attitudes. The data are shown below in Table 11.1.

Table 11.1: Data and Intermediate Statistics for Religion Questionnaire.

Student	Pre-test	Post-test	X_{DIFF}	$X_{DIFF}-M_{DIFF}$	$(X_{DIFF}-M_{DIFF})^2$
1	20	22	+2	0.44	0.1936
2	13	11	-2	-3.56	12.6736
3	4	8	+4	2.44	5.9536
4	10	17	+7	5.44	29.5936
5	16	12	-4	-5.56	30.9136
6	21	26	+5	3.44	11.8336
7	13	12	-1	-2.56	6.5536
8	14	16	+2	0.44	0.1936
9	12	13	+1	-0.56	0.3136

$\Sigma = 98.22$ $\Sigma X_{DIFF} = 14$

$M_{DIFF} = 1.56$ $\sigma^2_{DIFF} = \dfrac{98.22}{8} = 12.28$

We will again use our 5 step approach to determining statistical significance.

Step 1: Hypotheses. If there is no difference between pre-test and post-test scores, that means that the null hypothesis would be true. Therefore,

$$H_0: M_{DIFF} = 0$$

$$H_1: M_{DIFF} \neq 0$$

Step 2: Sampling Distribution. We are still using the t-distribution, with df = N-1. In our example, df = 9 - 1 = 8. Remember, N represents the number of pairs of observations we have, not the total number of observations when performing this type of t-test.

Step 3: Level of Significance. Setting α = .05, and with df = 8, our critical value for t (derived from the t distribution table in the back of your book) would be...

$$t'(8) = 2.306.$$

Step 4: Perform the Statistical Test.

$$t = \frac{M_{DIFF}}{\sigma_{MDIFF}}$$

$$\text{with } \sigma_{MDIFF} = \sqrt{\frac{\sigma^2}{N}}$$

In our example,

$$t = \frac{1.56}{\sigma_{MDIFF}}$$

$$\text{with } \sigma_{MDIFF} = \sqrt{\frac{12.278}{9}}$$

therefore,

$$t = \frac{1.56}{1.168} = 1.34$$

Step 5: Decision and Interpretation. Since our t value of 1.34 is less than our critical value of 2.306, we fail to reject the null hypothesis. We then can state that there is no statistically significant difference in the religious values scores before and after the religion course.

The official notation would be t(8) = 1.34, p>.05. This shows that the probability of the null hypothesis being true is greater than our alpha level of .05. Thus we must retain the null hypothesis.

ACTIVITY 11.1: The following scores were obtained before and after a sensitivity-training workshop. The scores measure how sensitive people are to the needs of others. Between pre- and post-measurements, we placed these individuals in the workshop in hopes of raising their awareness to the needs of others. Higher scores indicate greater sensitivity.

First, complete the intermediate statistics.

Participant	Pre-test	Post-test	X_{DIFF}	$X_{DIFF}-M_{DIFF}$	$(X_{DIFF}-M_{DIFF})^2$
1	4	6			
2	3	5			
3	7	9			
4	2	3			
5	6	7			

Now, complete the 5 step process to determine if the sensitivity workshop was successful.

1.   Hypotheses

2.   Sampling Distribution

3.   Level of Significance

4.   Perform the Statistical Test

5.   Decision and Interpretation

CONFIDENCE INTERVALS

The confidence interval formula takes the following form:

$$CI = M_{DIFF} \pm (t_{\alpha})(\sigma_{MDIFF}) = 1 - \alpha$$

For the 95% confidence interval in the religion example,

$$CI = 1.56 \pm (2.306)(1.168) = 1 - .05$$

$$CI = 1.56 \pm 2.69 = .95$$

$$CI = -1.13 \text{ to } 4.25$$

We can say that 95% of the time, we will expect to find a mean difference from pre-test to post-test of between -1.13 points and 4.25 points. It is within this range where we will fail to reject the null hypothesis. Since our mean difference was 1.56, this falls within the range of a non significant difference.

ACTIVITY 11.2: Based on the information from Activity 11.2, calculate the 95% confidence interval.

EFFECT SIZE

The effect size is similar to that of a t-test for a single mean.

$$d = \frac{M_{DIFF}}{\sqrt{\sigma^2}}$$

In our example,

$$d = \frac{1.556}{\sqrt{12.278}} \qquad = \frac{1.556}{3.504} = 0.44$$

Therefore, we have a moderate effect size.

ACTIVITY 11.3: Based on the information from Activity 11.1, calculate the effect size.

ADDITIONAL ACTIVITIES

1. A certain psychology professor was assigned 20 new freshman advisees. Before the start of the school year, he administered a test of conservatism-liberalism (pre-test). He then administered the same test again 4 years later to the same students (post-test). The higher the score, the more liberal the attitude. Summary statistics are below.

    $\Sigma X_{DIFF} = 35$

    $M_{DIFF} = 1.75$

    $\sigma^2_{DIFF} = 9.28$

a. Perform the statistical analysis to determine whether there is a statistically significant change over time.
b. Describe the results.
c. Calculate the 95% confidence interval
d. Calculate the effect size

2. A sports psychologist wants to investigate the effect of a particular relaxation technique versus no technique in a group of college basketball players. The ten basketball players attempted 30 foul shots, on two different days, after either participating in the relaxation technique or not. Summary statistics are below.

    $\Sigma X_{DIFF} = 62$

    $M_{DIFF} = 2.07$

    $\sigma^2_{DIFF} = 8.08$

a. Perform the statistical analysis to determine whether there is a statistically significant change in scores.
b. Describe the results.
c. Calculate the 95% confidence interval
d. Calculate the effect size

MODULE XII: ONE-BETWEEN ANALYSIS   OF   VARIANCE (ANOVA)

INTRODUCTION

The previous two modules dealt with situations where we wanted to determine whether there was a difference between either two independent or two related group means. This module and the next will take these analyses one step further. What if instead of having two means to compare, we have three, or four, or more? When it is the case that we have three or more means to compare, we move to a different type of statistical analysis known as the Analysis of Variance.

In this module, we will be covering the One-Between Analysis of Variance (ANOVA). The One-between ANOVA is used to test the null hypothesis that the means of J groups are all equal, where J indicates the actual number of groups that we will have. It is an extension of the independent groups t-test, in that we have three or more independent group means.

You may ask the question: "Why don't we just do multiple independent t-tests between all possible combinations of the groups?" First, this would be tedious and time consuming. Consider the situation where we have 5 groups, which would not be uncommon in behavioral science research. Perhaps we were interested in how well different athletic sport groups perform on a test of eye-hand coordination, and we composed our groups as follows:

Group 1:   Basketball
Group 2:   Baseball
Group 3:   Swimming
Group 4:   Track/Field
Group 5:   Volleyball

Perhaps we thought that athletes who typically engage in sports in which eye/hand coordination is very important (like basketball, baseball, and volleyball) would have better eye/hand coordination than those sports where such eye/hand coordination may be less important (like swimming and track/field). If we were to take a t-test approach to determining differences between means, we would have to perform the following:

Group 1 vs. Group 2
Group 1 vs. Group 3
Group 1 vs. Group 4
Group 1 vs. Group 5
Group 2 vs. Group 3
Group 2 vs. Group 4
Group 2 vs. Group 5
Group 3 vs. Group 4
Group 3 vs. Group 5
Group 4 vs. Group 5

Thus, we would have to perform 10 different t-tests, instead of just one ANOVA.

Second, there would be an overall inflated Type I Error rate. When we set our alpha level, we set it for each individual test that we perform. Therefore, if we choose $\alpha=.05$ and perform 10 different tests, our true $\alpha = .50$ (10 times $\alpha=.05$). That produces a 50% likelihood of making a Type I Error, which is unacceptable.

Finally, the use of analysis of variance allows for us to determine very complex patterns and interactions among variables that we would be unable to if we used a less sophisticated test (like a t-test). This will be most important when we start to integrate multiple independent variables into an experiment, rather than just one as will be done in this module.

So, instead of performing multiple t-tests, we first perform an ANOVA. The ANOVA will tell us whether or not there are any significant differences among any of our means. If there are no significant differences, we can stop right there and accept the null hypothesis. If there are significant differences, we will then follow-up our ANOVA with what are called post-hoc tests. These post-hoc tests will tell us exactly which means are significantly different from which other means.

LAYOUT OF THE ONE-BETWEEN ANOVA

The basic layout for the one-between ANOVA is shown in Table 12.1. Levels of the independent variable can be called treatments, conditions, groups, or cells.

Table 12.1: Basic Layout for a One-Between ANOVA.

Group 1	Group 2	Group 3	Group 4	...	Group J
$X_{1,1}$	$X_{1,2}$	$X_{1,3}$	$X_{1,4}$	...	$X_{1,J}$
$X_{2,1}$	$X_{2,2}$	$X_{2,3}$	$X_{2,4}$	...	$X_{2,J}$
$X_{3,1}$	$X_{3,2}$	$X_{3,3}$	$X_{3,4}$	...	$X_{3,J}$
$X_{4,1}$	$X_{4,2}$	$X_{4,3}$	$X_{4,4}$	...	$X_{4,J}$
.	.	.	.	...	.
.	.	.	.	...	.
.	.	.	.	...	.
$X_{n,1}$	$X_{n,2}$	$X_{n,3}$	$X_{n,4}$	...	$X_{n,J}$

In the above layout, we have presented a few new notations, which are important to note about the one-between ANOVA. In addition, there are some new summary notations with which you should become familiar. These include the following:

X = a data point. This is no different from how we have addressed a value of X in the past.

$X_{i,j}$ = the score of the ith observation in the jth group, with actual numbers being substituted for i and j once the data have been collected. Since we will have several observations, comprising a variety of independent groups, this notation will allow us to keep better track of our data. For example, if the value of the 5th observation in the 3rd group was 12, this would be indicated as: $12_{5,3}$.

J = the number of groups.

n_j = the total number of observations in a particular group. For example, if there were 25 people in group 4, this would be indicated as: $n_4 = 25$.

N = the total number of observations or data points in the entire data set.

M_j = mean of the jth group, with J indicating which particular group you are referring to.

M.. = the grand mean, indicating the mean of all scores combined in your data set.

NULL AND ALTERNATIVE HYPOTHESES

The null hypothesis for the ANOVA indicates that there are no significant difference among any of the group means. Thus, a typical form for the null hypothesis is

$$H_0: M_1 = M_2 = M_3 = M_4 = ... = M_J$$

However, contrary to the way we have looked at the alternative hypothesis in the past, we can not say the following

$$H_1: M_1 \neq M_2 \neq M_3 \neq M_4 \neq ... \neq M_J$$

The above alternative hypothesis implies that all group means (no matter how many groups we may have) will be significantly different from each other. That is not the goal of the analysis of variance. An ANOVA is designed to indicate whether at least one significant difference is present among the means. Therefore, the alternative hypothesis will be accepted even if only one mean differs from only one other mean.

Because of this, the alternative hypothesis is written as follows:

$$H_1: \ \Sigma(M_J - M..)^2 \neq 0$$

This implies that if you subtracted the mean of each group from the overall (grand) mean, squared that value for each individual group, and added up the squared values, that you would get something other than zero.

PARTITIONING THE SUM OF SQUARES AND VARIANCE

Consider the situation where you have four groups of people-- freshman, sophomores, juniors, and seniors. You are interested in knowing whether or not there are any statistically significant differences among these four groups in regards to school spirit. You administer a quick questionnaire concerning school spirit to a sample of each of the four matriculation groups. You would now like to perform some intermediate statistics to characterize your data.

The intermediate statistics you perform can be done in one of two ways. First, you might want to break the data down based on matriculation year, thus getting descriptive statistics (mean, variance, standard deviation, etc.) individually for freshmen, sophomores, juniors, and seniors. Second, you might want to lump all of the data together into one big data file to determine the descriptive statistics for your entire sample of students. By performing each of these different analyses, you have just set the foundation for the major underlying theoretical construct of ANOVA.

In an analysis of variance, we are ultimately interested in comparing two different types of variance: the variability within groups and the variability between groups. Therefore, the deviation of each individual score (X) from the grand mean can be expressed as the sum of two parts: the deviation of an individual score from the mean of the group it is in, and the deviation of the group mean from the grand mean.

In our example above on school spirit, we can imagine one particular person who filled out a questionnaire. We will call this person Pete. Pete is also a junior in college. How can we characterize the variance related to Pete's participation? First, we can see how Pete's school spirit is in relation to the other juniors. Second, we can see how the group of juniors' (of which Pete is a member) school spirit is in relation to the overall sample of all students. And that is what analysis of variance is all about. It is taking these two pieces of variance (within group and between group), and comparing them. How does that help us determine statistical significance? Let us hold off on that question until we actually learn how to calculate the different pieces of variance.

As you may recall, variance is equal to the sum of squares divided by the degrees of freedom. Therefore, the starting point for getting the variance in an ANOVA is to calculate the sum of squares. Since we eventually want to get the between variance and the within variance, there will be two sums of squares that need to be calculated. In addition, we can calculate the sum of squares for the entire sample (which would just be the addition of the between and within sums of squares).

$$SS_{TOTAL} = \Sigma\Sigma(X_{ij} - M..)^2 = \text{The sum of the squared deviations of each score from the grand mean.}$$

$$SS_{BETWEEN} = n\Sigma(M_J - M..)^2 = \text{The sum of the squared deviations of each group mean from the grand mean, weighted by } n.$$

$$SS_{WITHIN} = \Sigma\Sigma(X_{ij} - M_J)^2 = \text{The sum of the squared deviations of each score from the mean of the group it is in.}$$

Since variance is a sum of squares divided by its degrees of freedom, we can convert each of the above sums of squares into a variance estimate. An old-fashioned term for a variance estimate is "mean-square," short for mean squared deviation from the mean. This term is retained in the ANOVA, and we call our variance estimates mean squares (denoted as MS).

$$MS_{TOTAL} = \frac{SS_{TOTAL}}{df_{TOTAL}}$$

$$= \frac{\Sigma\Sigma(X_{ij} - M..)^2}{N-1}$$

$$MS_{BETWEEN} = \frac{SS_{BETWEEN}}{df_{BETWEEN}}$$

$$= \frac{n\sum(M_J - M..)^2}{J - 1}$$

$$MS_{WITHIN} = \frac{SS_{WITHIN}}{df_{WITHIN}}$$

$$= \frac{\sum\sum(X_{ij} - M_J)^2}{N - J}$$

THE F-TEST IN THE ANOVA

Now, back to the question of how these two pieces of variance allow us to determine statistical significance. The ANOVA involves comparing the ratio of $MS_{BETWEEN}$ to MS_{WITHIN}. MS_{WITHIN} is an estimator of general variance regardless of whether or not the null hypothesis is true. So you can think of it as how your observations will vary given that no experimental manipulation has been made. By experimental manipulation we could mean many things from giving participants some stimulus, assigning them to groups, grouping them based on some personal characteristic, etc. $MS_{BETWEEN}$ is an estimator of the MS_{WITHIN} plus the extent to which the means really differ due to some experimental manipulation. Therefore, if there are no differences due to your subjecting participants to experimental manipulations, the two mean squares are both estimating the exact same quantity. As a result, testing the null hypothesis of equal means is exactly equivalent to testing whether or not the two variance estimates are equal.

To do this, we calculate an F ratio (or F score), as follows:

$$F = \frac{MS_{BETWEEN}}{MS_{WITHIN}}$$

with (J - 1) and (N - J) df.

If there is no effect of an experimental manipulation, the $MS_{BETWEEN}$ and MS_{WITHIN} estimates will be exactly the same (or very close). This would produce an F score = 1. The more of an effect an experimental manipulation has had, the larger the F score will be. At some point the F score will become large enough for us to make the statement that such an F score is unlikely to occur just by chance. When that happens, we will reject the null hypothesis.

AN EXPERIMENTAL EXAMPLE

Imagine that we wish to study the effects of an experimental drug designed to alleviate symptoms of depression. Thirty patients are selected and then assigned at random to one of three groups. The Placebo Control group is given an inert substance, the Low Dose Drug group is given 10 mg of the drug, and the High Dose Drug group is given 50 mg of the drug. The dependent variable is a measure of how happy our patients feel, with higher scores indicating greater happiness. The data table and some of the intermediate calculations are shown in Table 12.2.

SS_{TOTAL} is ordinarily calculated first. From Table 12.2, we see that the grand mean is 12.6. SS_{TOTAL} represents the variability of all 30 scores around this mean.

$$SS_{TOTAL} = \Sigma\Sigma(X_{ij} - M..)^2$$

$$= \Sigma\Sigma(X_{ij} - 12.6)^2$$

$$\begin{aligned}
= &(10 - 12.6)^2 + (7 - 12.6)^2 + (9 - 12.6)^2 + (8 - 12.6)^2 + \\
&(15 - 12.6)^2 + (3 - 12.6)^2 + (8 - 12.6)^2 + (9 - 12.6)^2 + \\
&(11 - 12.6)^2 + (9 - 12.6)^2 + (19 - 12.6)^2 + (12 - 12.6)^2 + \\
&(16 - 12.6)^2 + (14 - 12.6)^2 + (7 - 12.6)^2 + (8 - 12.6)^2 + \\
&(13 - 12.6)^2 + (10 - 12.6)^2 + (19 - 12.6)^2 + (9 - 12.6)^2 + \\
&(23 - 12.6)^2 + (14 - 12.6)^2 + (16 - 12.6)^2 + (18 - 12.6)^2 + \\
&(12 - 12.6)^2 + (13 - 12.6)^2 + (16 - 12.6)^2 + (17 - 12.6)^2 + \\
&(19 - 12.6)^2 + (14 - 12.6)^2
\end{aligned}$$

$$\begin{aligned}
= &6.76 + 31.36 + 12.96 + 21.16 + 5.76 + 92.19 + 21.16 \\
&+ 12.96 + 2.56 + 12.96 + 40.96 + .36 + 11.56 + 1.96 + \\
&31.36 + 21.16 + .16 + 6.76 + 40.96 + 12.96 + 108.16 + \\
&1.96 + 11.56 + 29.16 + .36 + .16 + 11.56 + 19.36 + \\
&40.96 + 1.96
\end{aligned}$$

$$= 613.2$$

Table 12.2:   Summary of Computations for the ANOVA Drug Example.

	Placebo	Low Dose	High Dose	
	10	19	23	
	7	12	14	
	9	16	16	
	8	14	18	
	15	7	12	
	3	8	13	
	8	13	16	
	9	10	17	
	11	19	19	
	9	9	14	
n	10	10	10	$N = 30$
ΣX	89	127	162	$\Sigma\Sigma X = 378$
M	8.9	12.7	16.2	$M.. = 12.6$
$\Sigma(X-M)^2$	82.9	168.1	95.6	$\Sigma\Sigma(X-M)^2 = 346.6$
σ^2_J	9.21	18.68	10.62	

$SS_{BETWEEN}$ is computed from the J group means (8.9, 12.7 and 16.2) as shown below...

$$SS_{BETWEEN} = n\Sigma(M_J - M..)^2$$

$$= 10\Sigma(M_J - 12.6)^2$$

$$= 10[(8.9 - 12.6)^2 + (12.7 - 12.6)^2 + (16.2 - 12.6)^2]$$

$$= 10(13.69 + .01 + 12.96)$$

$$= 10(26.66)$$

$$= 266.6$$

Note that when calculating $SS_{BETWEEN}$ we use n rather than N.

Finally, SS_{WITHIN} is calculated using each score in relation to the group it is in.

$SS_{WITHIN} = \Sigma\Sigma(X_{ij} - M_J)^2$

$$= (10 - 8.9)^2 + (7 - 8.9)^2 + (9 - 8.9)^2 + (8 - 8.9)^2 +$$
$$(15 - 8.9)^2 + (3 - 8.9)^2 + (8 - 8.9)^2 + (9 - 8.9)^2 +$$
$$(11 - 8.9)^2 + (9 - 8.9)^2 + (19 - 12.7)^2 + (12 - 12.7)^2 +$$
$$(16 - 12.7)^2 + (14 - 12.7)^2 + (7 - 12.7)^2 + (8 - 12.7)^2 +$$
$$(13 - 12.7)^2 + (10 - 12.7)^2 + (19 - 12.7)^2 + (9 - 12.7)^2 +$$
$$(23 - 16.2)^2 + (14 - 16.2)^2 + (16 - 16.2)^2 + (18 - 16.2)^2$$
$$+ (12 - 16.2)^2 + (13 - 16.2)^2 + (16 - 16.2)^2 + (17 - 16.2)^2$$
$$+ (19 - 16.2)^2 + (14 - 16.2)^2$$

$$= 1.21 + 3.61 + .01 + .81 + 37.21 + 34.81 + .81 + .01 +$$
$$4.41 + .01 + 39.69 + .49 + 10.89 + 1.69 + 32.49 + 22.09$$
$$+ .09 + 7.29 + 39.69 + 13.69 + 46.24 + 4.84 + .04 + 3.24$$
$$+ 17.64 + 10.24 + .04 + .64 + 7.84 + 4.84$$

$$= 346.6$$

Since $SS_{WITHIN} + SS_{BETWEEN}$ must be equal to SS_{TOTAL}, it is easy to check your work to make sure you did not make any errors in the calculations.

THE ANOVA TABLE

Now we can begin to build what is called an ANOVA table. We begin by entering the sums of squares just calculated.

ANOVA Table for Drug Dosage Differences in Happiness

Source	SS	df	MS	F	p
Total	613.2				
Between	266.6				
Within	346.6				

Next, we enter the degrees of freedom for each of the sources of variance, remembering that...

$df_{TOTAL} = N-1$
$df_{BETWEEN} = J-1$
$df_{WITHIN} = N-J$

ANOVA Table for Drug Dosage Differences in Happiness

Source	SS	df	MS	F	p
Total	613.2	29			
Between	266.6	2			
Within	346.6	27			

Notice that $df_{TOTAL} = df_{BETWEEN} + df_{WITHIN}$.

Next we must calculate the mean square. Since mean square is an indication of variance, and variance is calculated by dividing the sum of squares by the degrees of freedom, we can do this for our between and within sources of variance.

ANOVA Table for Drug Dosage Differences in Happiness

Source	SS	df	MS	F	p
Total	613.2	29			
Between	266.6	2	133.3		
Within	346.6	27	12.84		

Our final step is obtain a critical value for F and to calculate our actual F score. To obtain the critical value, we will be using a new table known as the F distribution (located in the back of your manual). The degrees of freedom for the numerator are the same as the degrees of freedom for our between source of variability, and the degrees of freedom for the denominator are the same as the degrees of freedom for our within source of variability.

In this example, we go across the top line of the F distribution table until we find 2 (our between degrees of freedom) and along the left side until we find 27 (our within degrees of freedom). With $\alpha=.05$ (lightface type) and 2 and 27 degrees of freedom, we see that the critical value of F is 3.35, so $F'(2,27) = 3.35$. If the score we get for our F test is 3.35 or greater, we will say there is a significant difference somewhere among the means of our 3 groups, and we will reject the null hypothesis. If the score we get for our F test is less than 3.35, we say there is no significant difference among any of the means of our groups, and accept the null hypothesis.

The actual F value is calculated by dividing $MS_{BETWEEN}$ by MS_{WITHIN}. In our example, we get 133.3/12.84 = 10.38. Since 10.38 is greater than our critical value of 3.35, we reject the null hypothesis and can say that at least one pair of group means is significantly different. The proper notation for the F test would be $F(2,27) = 10.38$, $p<.05$. Our final ANOVA table would take on the following appearance:

ANOVA Table for Drug Dosage Differences in Happiness

Source	SS	df	MS	F	p
Total	613.2	29			
Between	266.6	2	133.3	10.38	<.05
Within	346.6	27	12.84		

It is also customary to present either a table of means and standard deviations, or a graph, or both, as in Table 12.3 and Figure 12.1.

Table 12.3: Happiness Means and Variances by Drug Dose Condition.

	Placebo	Low Dose	High Dose
Mean	8.9	12.7	16.2
Variance	9.2	18.7	10.6

Figure 12.1: Bar Graph of Happiness Means by Drug Dose Condition

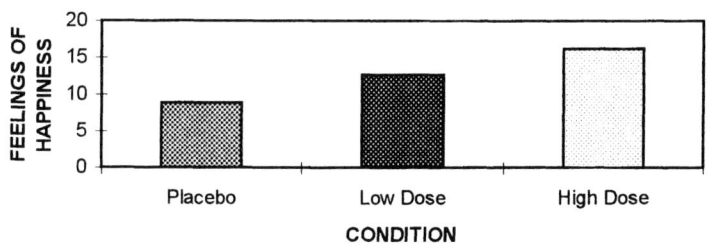

Means for Anxiety Measure by Dosage Condition

We have now determined that there is at least one significant difference among the means of our groups. But where is/are those significant differences? In order to determine where they are, we need to perform what is called a post hoc test.

POST HOC TESTS AND COMPARISONS

The ANOVA is designed to test whether or not there is a statistically significant difference between at least one pair of means. However, it does not give us any information concerning how many pairs of means are statistically significant, or which pairs they are. In order to determine where the statistically significant differences are, we would have to make comparisons among the individual group means. To do this, we use what is called a post-hoc (ie., after the fact) test. There are several types of post-hoc tests available, depending on how much you wish to control Type I Error, how many comparisons you will be performing, how much power you want to retain, etc. Such decisions will be left for more in-depth studies, however, and we will be choosing to employ the popular Tukey HSD Procedure.

THE TUKEY HSD PROCEDURE

The Tukey HSD (Honestly Significant Difference) procedure is a post hoc test that is neither too conservative nor too liberal in regards to controlling Type I Error. It is calculated as follows:

$$q = \frac{M_J - M_{J'}}{\sqrt{\dfrac{MS_{WITHIN}}{n}}}$$

with df = J and (N-J)

M_J and $M_{J'}$ represent two group means that you are comparing, and the test is called a q- test.

Make note that the above equation is appropriate when we have equal n's in each group. When the n's of each group are not equal, then the denominator of the Tukey procedure is the following:

$$\sqrt{\frac{MS_{WITHIN}}{2} * \frac{1}{n_1} + \frac{1}{n_2}}$$

Just like with the t and F scores, we need to find a critical value of q, or q'. Also, just like our previous statistical analyses we will compare the actual score we receive after performing the q test to the q' value. If the q value is equal to or greater than q', then there is a statistically significant difference between those two pairs of means. If the q value is less than q', then there is not a significant difference between those two pairs of means. A table of critical values for q (Tukey HSD) can be found in the back of your manual.

For our drug dosage and happiness example, we will be working with J=3 and (N-J)=27 degrees of freedom. With α=.05, and 3 and 24 degrees of freedom (since there is no 27 on our table, we use the next lowest number), we find that the critical value of q is 3.52, so q'(3,24)=3.52. When we derive our actual q values, any value equal to or greater than 3.52 will be considered statistically significant. Any q value lower than 3.52 will not be considered statistically significant.

Since the denominator of the q test statistic is constant for all comparisons, we can calculate it first, as follows:

$$\sqrt{\frac{MS_{WITHIN}}{n}} = \sqrt{\frac{12.84}{10}} = \sqrt{1.28} = 1.13$$

Since we have 3 groups, we want to make the following comparisons:

Placebo (M=8.9) vs. Low Dose (M=12.7)

Placebo (M=8.9) vs. High Dose (M=16.2)

Low Dose (M=12.7) vs. High Dose (M=16.2)

The q scores for each of these comparisons would be calculated as follows:

$$\text{Placebo vs. Low Dose} = \frac{(8.9 - 12.7)}{1.13} = -3.36$$

$$\text{Placebo vs. High Dose} = \frac{(8.9 - 16.2)}{1.13} = -6.46$$

$$\text{Low Dose vs. High Dose} = \frac{(12.7 - 16.2)}{1.13} = -3.10$$

We take the absolute values when determining significance, so we are comparing 3.36, 6.46 and 3.10 against the q critical value of 3.52. We can see that only one q value is greater than the q critical value-- 6.46, which is the comparison of the Placebo vs. High Dose.

So what conclusions can we draw? We can say that the high dose of 50mg is significantly better in promoting greater feelings of happiness in our depressed patients in comparison to a placebo, that the low dose of 10mg is not different from a placebo, and that there is no difference between the 10mg dose and the 50mg dose.

ACTIVITY 12.1: We are interested in whether or not there are differences in the motivation to succeed in college among three different college majors. We administer a test of motivation to groups of Psychology, Business, and Chemistry Majors. The higher the score, the more motivated the individuals are to succeed. Summary data are presented below:

Psychology	Business	Chemistry
8	9	8
7	10	10
10	9	10
8	10	7
6	7	5
9	9	7
10	8	6
8	10	7

n	8	8	8	N = 24
ΣX	66	72	60	$\Sigma\Sigma X = 198$
M	8.25	9.0	7.5	M.. = 8.25
$\Sigma(X-M)^2$ 13.5		8.0	22	$\Sigma\Sigma(X-M)^2 = 43.5$
σ^2_J	1.93	1.14	3.14	

Calculate the appropriate F test to determine whether or not there are significant differences among the three groups. If there are significant differences, perform the appropriate post-hoc analysis to determine where these differences are.

$$SS_{TOTAL} = \Sigma\Sigma(X_{ij} - M..)^2$$

$$SS_{BETWEEN} = n\Sigma(M_J - M..)^2$$

$$SS_{WITHIN} = \Sigma\Sigma(X_{ij} - M_J)^2$$

ANOVA TABLE

Source	SS	df	MS	F	p
Total					
Between					
Within					

What conclusions can you draw so far?

Perform the Post-hoc analyses (if appropriate)

What are your final conclusions and interpretation?

ADDITIONAL ACTIVITIES

1. In a study on the effects of Pavlovian conditioning, three groups of rats were presented a tone and rat chow. The tone began three seconds before the rat chow was presented, at the same time the rat chow was presented, or three seconds after the rat chow was presented. The dependent variable was the number of drops of saliva elicited by the tone alone after 100 pairings of the tone and food powder. There were 20 rats in each group.

a. Fill in the blanks with the appropriate numbers, setting $\alpha=.05$.

Condition	Mean Drops of Saliva
3 seconds before	27
simultaneous	14
3 seconds after	3

Source	SS	df	MS	F	p
Total	3996				
Between	_____	__	_____	_____	_____
Within	910	__	_____		

b. Perform the appropriate post hoc tests.
c. Describe the outcome of the experiment.

2. A psychopharmacologist conducted a study on the effects of lithium for the treatment of depression. Twenty-seven patients were chosen whose principal problem was depression. There were 3 groups: placebo, low dose, and high dose. There were 9 patients in each group. At the end of 30 days, the patients were evaluated for their level of depression. Higher scores indicate higher levels of depression.

a. Fill in the blanks with the appropriate numbers, setting $\alpha=.05$.

Group	Depression Score
Placebo	13.2
Low Dose	13.2
High Dose	8.3

Source	SS	df	MS	F	p
Total	____	__			
Between	143.41	__	____	____	____
Within	365.11	__	____		

b. Perform the appropriate post hoc tests.
c. Describe the outcome of the experiment.

3. A human rights advocate, concerned about the low proportion of female public-school administrators, prepared a questionnaire to measure attitudes about hiring a woman to fill the vacancy of superintendent of schools in a large city. The higher the score, the more positive the attitude. The questionnaire was administered to a random sample of students, parents, and teachers. Analyze the data, conduct any additional analyses that you feel are necessary, and write a statement about attitudes toward hiring a female superintendent.

Students	Parents	Teachers
20	16	10
18	14	9
16	10	7
14	9	7
12	7	5
25	21	14
23	20	13
19	20	11
19	18	10
15	14	9

MODULE XIII: ONE-WITHIN ANALYSIS OF VARIANCE (ANOVA)

INTRODUCTION

We were able to draw a connection between the independent t-test and the One-Between ANOVA. We can also draw a connection between the related means t-test and the One-Within ANOVA. One of the ways in which we could use the related means t-test was in assessing whether there was a change in scores from one point in time (pre-test) to another point in time (post-test). The One-Within ANOVA will allow us to assess changes in scores that are taken three or more times.

We begin by having some variable of interest that we will measure several times. Each of these measurement times is known as a level, and we designate the number of levels by the letter J. Therefore, if we measured our variable four times, J=4. We only have one sample of n participants, and each participant will be measured under all levels of J. The general layout of a One-Within ANOVA is found in Table 13.1.

In the layout, we have presented a few new notations, which are important to note about the one-within ANOVA. In addition, there are some new summary notations with which you should become familiar. These include the following:

Table 13.1: General Layout of a One-Within ANOVA.

Participant	Level 1	Level 2	Level 3	...	Level J
1	$X_{1,1}$	$X_{1,2}$	$X_{1,3}$	...	$X_{1,J}$
2	$X_{2,1}$	$X_{2,2}$	$X_{2,3}$	...	$X_{2,J}$
3	$X_{3,1}$	$X_{3,2}$	$X_{3,3}$	...	$X_{3,J}$
4	$X_{4,1}$	$X_{4,2}$	$X_{4,3}$	...	$X_{4,J}$
.	.	.	.	...	.
.	.	.	.	...	.
.	.	.	.	...	.
n	$X_{n,1}$	$X_{n,2}$	$X_{n,3}$	...	$X_{n,J}$

X = a data point. This is no different from how we have addressed a value of X in the past.

$X_{i,j}$ = the score of the ith person in the jth level, with actual numbers being substituted for i and j once the data have been collected. For example, if the value of the 5th participant in the 3rd level was 12, this would be indicated as: $12_{5,3}$.

J = the number of levels.

n_j = the total number of observations in a particular level. Note, however, that all levels should have the same number of observations, since this is a repeated measured design among the same participants.

N = the total number of observations in the entire experiment.

M_j = mean of the jth level, with J indicating which particular level you are referring to.

M.. = the grand mean, indicating the mean of all scores combined in your
      data set.

Because we are measuring the same people at more than one time,
we can assess variability due to individual differences among participants.
By doing this, we are able to account for more variability, and thus increase
the likelihood of noting a significant difference if one exists.

PARTITIONING THE SUM OF SQUARES

SS_{TOTAL} is computed exactly as in the one-between ANOVA.

$$SS_{TOTAL} = \sum\sum(X_{ij} - M..)^2$$

In the one-between ANOVA, SS_{TOTAL} was partitioned into two
parts: $SS_{BETWEEN}$ and SS_{WITHIN}. In the one-within ANOVA, SS_{TOTAL} is
partitioned into three parts: $SS_{TREATMENTS}$, $SS_{SUBJECTS}$, and SS_{ERROR}. It is
therefore the case that SS_{WITHIN} for the one-between design is itself broken
into two parts ($SS_{SUBJECTS}$ and SS_{ERROR}) in the one-within design.

$$SS_{TREATMENTS} = n\sum(M_J - M..)^2$$

$SS_{TREATMENTS}$ is also computed in exactly the same way as $SS_{BETWEEN}$
was computed in the one-between design. This piece of the variability is
referred to as "treatments," and represents the variability of the treatment
means from the grand mean.

$$SS_{SUBJECTS} = J\sum(M_i - M.)^2$$

$$SS_{ERROR} = \sum\sum(X_{ij} - M_J - M_i + M..)^2$$

The subjects value relates to individual differences among the
participants in the study, while the error value relates to any residual
variability that has not been accounted for by either our treatments or the
participant individual differences. Being able to account for these individual
differences is one of the greatest benefits of using a within design. We were
unable to do this in a between design, since each person was only measured

once (thus the idea of independent groups).

MEAN SQUARES AND DEGREES OF FREEDOM

We use a similar format to calculate the mean squares in the one-within ANOVA as we did in the one-between ANOVA. Mean square is just another word for variance, and variance is equal to the sum of squares divided by the degrees of freedom.

$$MS_{TOTAL} = \frac{SS_{TOTAL}}{df_{TOTAL}}$$

$$= \frac{\sum\sum(X_{ij} - M..)^2}{N - 1}$$

$$MS_{TREATMENTS} = \frac{SS_{TREATMENTS}}{df_{TREATMENTS}}$$

$$= \frac{n\sum(M_J - M..)^2}{J - 1}$$

$$MS_{SUBJECTS} = \frac{SS_{SUBJECTS}}{df_{SUBJECTS}}$$

$$= \frac{J\sum(M_i - M..)^2}{n - 1}$$

$$MS_{ERROR} = \frac{SS_{RESIDUAL}}{df_{RESIDUAL}}$$

$$= \frac{\sum\sum(X_{ij} - M_J - M_i + M..)^2}{(n - 1)(J - 1)}$$

$$= \frac{SS_{TOTAL} - SS_{TREATMENTS} - SS_{SUBJECTS}}{(n - 1)(J - 1)}$$

AN EXPERIMENTAL EXAMPLE

Suppose that we are the director of a large corporation, and we want to see what the public opinion is of our company. To assess public opinion, we obtain a sample of consumers and ask them to complete a questionnaire. We then send them some free samples of our product...let's say we make soap, and we send our sample of participants a free bar of soap. After they have had a chance to receive our free sample, we again assess what their opinion is of our company. If we stopped right there, we could use a related means t-test to assess change in company opinion. However, let's say we would like to follow-up at some point later in time, and we assess people's opinion of the company for a third time 6 months later. We now have the same group of people being measured on 3 different occasions, which calls for a one-within ANOVA.

The data and some intermediate statistics are shown in Table 13.2. Higher scores represent a more positive view of the company.

Table 13.2: Summary of Computations for the ANOVA Company Public Opinion Example.

Participant	Initial Assessment	Assessment After Free Sample	Assessment 6 months later	Σ	M_i
1	10	28	21	59	19.67
2	13	27	24	64	21.33
3	17	36	26	79	26.33
4	18	31	20	69	23.00
5	15	33	26	74	24.67
6	18	32	29	79	26.33
7	22	41	31	94	31.33
ΣX	113	228	177	$\Sigma\Sigma X = 518$	
M_j	16.14	32.57	25.29	$M.. = 24.67$	
$n = 7$	$N = 21$	$J = 3$			

The sums of squares are calculated as follows:

$$SS_{TOTAL} = \sum\sum(X_{ij} - M..)^2$$

$$
\begin{aligned}
&= (10 - 24.67)^2 + (13 - 24.67)^2 + (17 - 24.67)^2 + \\
&\quad (18 - 24.67)^2 + (15 - 24.67)^2 + (18 - 24.67)^2 + \\
&\quad (22 - 24.67)^2 + (28 - 24.67)^2 + (27 - 24.67)^2 + \\
&\quad (36 - 24.67)^2 + (31 - 24.67)^2 + (33 - 24.67)^2 + \\
&\quad (32 - 24.67)^2 + (41 - 24.67)^2 + (21 - 24.67)^2 + \\
&\quad (24 - 24.67)^2 + (26 - 24.67)^2 + (20 - 24.67)^2 + \\
&\quad (26 - 24.67)^2 + (29 - 24.67)^2 + (31 - 24.67)^2
\end{aligned}
$$

$$
\begin{aligned}
&= 215.21 + 136.19 + 58.83 + 44.49 + 93.51 + \\
&\quad 44.49 + 7.13 + 11.09 + 5.43 + 128.37 + 40.07 \\
&\quad + 69.39 + 53.73 + 266.67 + 13.47 + .45 + 1.77 \\
&\quad + 21.81 + 1.77 + 18.75 + 40.07
\end{aligned}
$$

$$= 1272.69$$

$$SS_{TREATMENTS} = n\sum(M_J - M..)^2$$

$$= 7[(16.14 - 24.67)^2 + (32.57 - 24.67)^2 + (25.29 - 24.67)^2$$

$$= 7[72.76 + 62.41 + .38]$$

$$= 7[135.55]$$

$$= 948.85$$

$$SS_{SUBJECTS} = J\sum(M_i - M..)^2$$

$$
\begin{aligned}
&= 3[(19.67 - 24.67)^2 + (21.33 - 24.67)^2 + (26.33 - 24.67)^2 \\
&\quad + (23 - 24.67)^2 + (24.67 - 24.67)^2 + (26.33 - 24.67)^2 + \\
&\quad (31.33 - 24.67)^2
\end{aligned}
$$

$$= 3[25 + 11.16 + 2.76 + 2.79 + 0 + 2.76 + 44.36]$$

$$= 3[88.83]$$

$$= 266.49$$

SS_{ERROR} $= SS_{TOTAL} - SS_{SUBJECTS} - SS_{TREATMENTS}$

$= 1272.69 - 266.49 - 948.85$

$= 57.35$

Notice that about 21% ($266.49 \div 1272.69$) of SS_{TOTAL} is accounted for by $SS_{SUBJECTS}$. Were we unable to account for this variance, it would have gone into the SS_{ERROR} term, thus increasing the error variance and decreasing the likelihood of finding a significant effect should one exist.

The beginnings of the ANOVA table are shown below.

ANOVA Table for Company Public Opinion

Source	SS	df	MS	F	p
Total	1276.69				
Subjects	266.49				
Treatments	948.85				
Error	57.35				

The next step is the addition of the degrees of freedom, remembering that...

$df_{TOTAL} = N-1$

$df_{SUBJECTS} = n-1$

$df_{TREATMENTS} = J-1$

$df_{ERROR} = (n-1)(J-1)$

ANOVA Table for Company Public Opinion

Source	SS	df	MS	F	p
Total	1276.69	20			
Subjects	266.49	6			
Treatments	948.85	2			
Error	57.35	12			

Next we calculate the mean square by dividing the sum of squares by the degrees of freedom. For the final F score, we are interested only in the treatments and the error values.

ANOVA Table for Company Public Opinion

Source	SS	df	MS	F	p
Total	1276.69	20			
Subjects	266.49	6			
Treatments	948.85	2	474.43		
Error	57.35	12	4.78		

Our final step is to obtain a critical value for F and to calculate the actual F score. To obtain the critical value, we again use the F distribution (found in the back of your manual). The degrees of freedom for the numerator are the same as the degrees of freedom for our treatments source of variability, whereas the degrees of freedom for the denominator are the same as the degrees of freedom for our error source of variability. In this example, we go across the top line until we find 2 and along the left side until we find 12. With $\alpha = .05$ and 2 and 12 degrees of freedom, we see that the critical value of F is 3.88, so $F'(2,12) = 3.88$. If the score from the F-test is 3.88 or greater, we will say that there is a significant difference somewhere among the means of our 3 treatments/conditions, and that we will reject the null hypothesis. If the score from the F-test is less than 3.88, we will say that there is no significant difference among any of the means of our treatments/conditions, and thus accept the null hypothesis.

The F value is calculated by dividing $MS_{TREATMENTS}$ by MS_{ERROR}. In our example, we get $474.43 \div 4.78 = 99.25$. Since 99.25 is greater than the critical value of 3.88, we reject the null hypothesis and say that at least one pair of means is significantly different. The final ANOVA table is below:

ANOVA Table for Company Public Opinion

Source	SS	df	MS	F	p
Total	1276.69	20			
Subjects	266.49	6			
Treatments	948.85	2	474.43	99.25	<.05
Error	57.35	12	4.78		

The results are presented graphically in Figure 13.1.

Figure 13.1: Bar Graph of Means by Treatment Condition for the Soap Company Example.

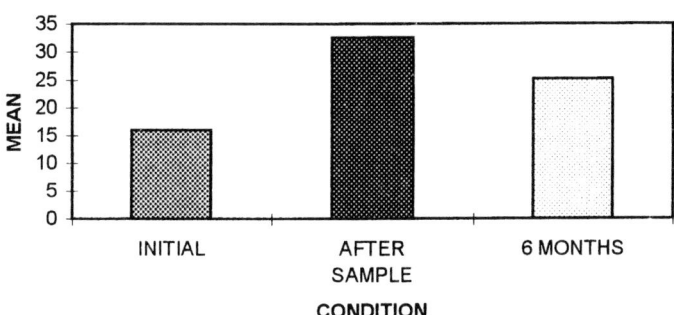

Means by Treatment Condition

Just like in the One-Between ANOVA, we can say that, based on our F-test, there is at least one statistically significant difference among the three assessment times. To find out where these differences are, we must again perform a post hoc test.

POST HOC COMPARISONS

For the reasons explained in Module 12, we will again be using the Tukey HSD as our post hoc test of choice. The test is calculated in much the same way as before:

$$q = \frac{M_J - M_J'}{\sqrt{\dfrac{MS_{ERROR}}{n}}}$$

with degrees of freedom = J and (N - J).

We begin by finding a critical value of q. A table of critical values can be found in the back of your manual. For our example, we will be working with J=3 and N-J=18 degrees of freedom (remember that N=21). With $\alpha=.05$ and 3 and 18 degrees of freedom, we find that the critical value of q is 3.61, so q'(3,18)=3.61. When we derive our actual q values, any value equal to or greater than 3.61 will be considered statistically significant. Any q value less than 3.61 will not be considered statistically significant.

Since the denominator of the q test statistic is constant for all comparisons, we can calculate it first, as follows:

$$\sqrt{\frac{MS_{ERROR}}{n}} = \sqrt{\frac{4.78}{7}} = \sqrt{.683} = .826$$

Since there are 3 measurement times (J=3), we want to make the following comparisons:

Initial Assessment (M=16.14) vs. After Sample (M=32.57)

Initial Assessment (M=16.14) vs. 6 Month Follow-up (M=25.29)

After Sample (M=32.57) vs. 6 Month Follow-up (M=25.29)

The q scores for each of these comparisons would be calculated as follows:

Initial Assessment vs. After Sample = $\dfrac{(16.14 - 32.57)}{.826}$ = -19.89

Initial Assessment vs. 6 Months = $\dfrac{(16.14 - 25.29)}{.826}$ = -11.08

After Sample vs. 6 Months = $\dfrac{(32.57 - 25.29)}{.826}$ = 8.81

We take the absolute values when determining significance, so we are comparing 19.89, 11.08 and 8.81 against the q critical value of 3.61. We can see that each of our comparisons is greater than the q critical value.

What conclusions can we draw? We can say that the impression of the company increased significantly after the free sample, then significantly decreased again after 6 months. However, after 6 months the impression of the company was still better than it was before the free sample was given.

ACTIVITY 13.1: Imagine we are interested in whether or not there are differences in athlete's mood when they are presented with different odors while they are working out. We administer a test of mood to a group of athletes while they run on a treadmill for 15 minutes. They complete 3 sessions, one while smelling peppermint, one while smelling jasmine, and one while smelling lemon. The higher the score, the better their mood. Summary data are presented below:

Athlete	Peppermint	Jasmine	Lemon	Σ	M_i
1	9	7	6	22	7.33
2	10	6	6	22	7.33
3	9	7	7	23	7.67
4	8	6	7	21	7.00
5	7	5	6	18	6.00
ΣX	43	31	32	$\Sigma\Sigma X = 106$	
M_j	8.6	6.2	6.4	$M.. = 7.07$	
n = 5	N = 15	J=3			

Calculate the appropriate F test to determine whether or not there are significant differences among the three treatment conditions.

$$SS_{TOTAL} = \Sigma\Sigma(X_{ij} - M..)^2$$

$$SS_{TREATMENTS} \quad = n\sum(M_j - M..)^2$$

$$SS_{SUBJECTS} \quad = J\sum(M_i - M_.)^2$$

$$SS_{ERROR} = SS_{TOTAL} - SS_{SUBJECTS} - SS_{TREATMENTS}$$

Source	SS	df	MS	F	p
Total					
Subjects					
Treatments					
Error					

What conclusions can be drawn thus far?

Post-hoc analyses (if appropriate)

What are your final conclusions and interpretation?

ADDITIONAL ACTIVITIES

1. Developmental psychologists study children at different ages. In this study, 12 children were studied three times, at ages 10, 11 and 12. At each age, they worked a set of 20 logic problems. The scores below represent the mean percentage correct on the set of 20 problems at different ages.

    Age 10 = 26.50
    Age 11 = 47.00
    Age 12 = 60.75

a. Complete the following ANOVA table.

Source	SS	df	MS	F	p
Total	42667	___			
Subjects	17506	___			
Age	_____	___	_____	_____	_____
Error	4038.67	___	_____		

b. Perform the appropriate post-hoc analyses.
c. Describe your results.

2. In another developmental psychology experiment, the children were studied three times, at ages 5, 6, and 7. At each age, they worked a set of 25 problems. Each problem was a measure of the child's understanding of the concept of conservation (that quantity remains the same despite changes in shape). The scores below represent percentage correct on the set of 25 problems.

Child	Age 5	Age 6	Age 7
1	4	12	68
2	12	24	80
3	20	16	88
4	20	20	84

a.   Perform a one-between ANOVA.
b.   Construct the ANOVA table.
c.   Perform the appropriate post-hoc analysis.
d.   Describe your results.

MODULE XIV: TWO-BETWEEN ANALYSIS OF VARIANCE (ANOVA)

INTRODUCTION

After having completed Modules XII and XIII, you should have a good understanding of what the analysis of variance can be used for. We have discussed the two basic ANOVA situations: one-between ANOVA and one-within ANOVA. However, what if you wanted to have two independent variables in a study and wished to assess their combined effect on some dependent variable? Or what if you wanted to have one independent variable be a between variable and the other independent variable be a within variable? Or what if you wanted to have four between variables and three within variables?

Any of the above combinations are easily handled through the use of an ANOVA. And the number of combinations is limited only to your imagination and research design (although interpretation becomes exponentially more difficult as you add new variables to be analyzed). Since the possible combinations are vast, there is no way to address each type of analysis. However, in this module we will examine the two-between analysis of variance. This is the situation where you have two independent variables that are both between-subjects in nature (thus the name).

There are many advantages to using a research design which makes use of more than one variable.

- **Efficiency.** One of the reasons we chose to perform a lower order ANOVA (like a one-between or one-within) instead of multiple t-tests was that the ANOVA is much more efficient. The same holds true for the use of multiple independent variables. The more variables you add to the analysis, the more efficient you will be as a statistician.

- **Power.** When we first discussed the use of a regression analysis, you were presented with the idea of "proportion of variance accounted for" and "proportion of variance not accounted for." Why were we unable to account for 100% of the variance? Of course, there are hundreds or thousands of possible influences on a dependent variable, but because of time constraints, money constraints, personnel constraints, etc., we are unable to measure each and every one. However, the more variables we are able to measure, the more variance we will be able to account for. The same holds true with using more advanced level, multi-variable ANOVAs. If we increase the number of independent variables, we will decrease the amount of uncontrolled variance, and therefore have a more powerful test.

- **Interaction effects.** An interaction effect is when the levels of two or more independent variables combine in such a way as to produce a change in the dependent variable, above and beyond the individual variable effects. Don't be worried about that definition...the example should make it clear.

Take a look at Figure 14.1. Imagine that we are interested in whether there are differences on a particular test of cognitive ability. Figure 14.1 shows the means of our participants' scores. We have two independent variables: sex of the participant and whether they go to a public or private school. We also, therefore, have four independent groups, with each group being labeled by a particular bar in the graph.

If the only variable we collected was male vs. female, what is the likelihood of finding a statistically significant difference? If we compute the mean for all males ($M=50$) and compare it to the mean for all females ($M=50$), we see that the means are exactly alike. Therefore, there would be

Figure 14.1: **Scores on a Test of Cognitive Ability for Males and Females in Public and Private Schools.**

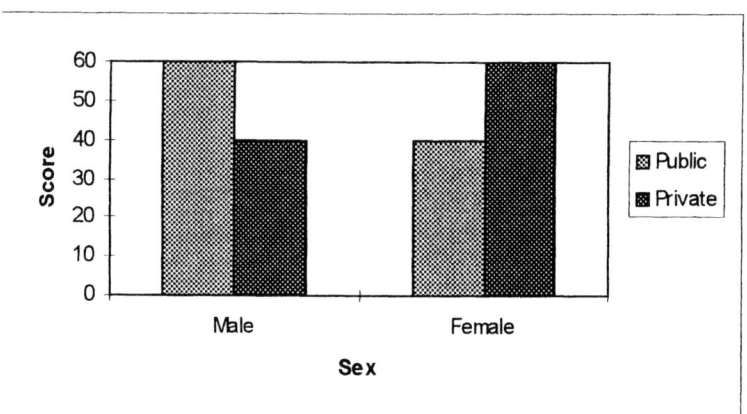

no significant difference. If the only variable we collected was public school versus private school, what is the likelihood of finding a statistically significant difference? If we compute the mean for all public school participants (M=50) and compare it to the mean for all private school participants (M=50), we again see that the means are exactly alike. This would also lead to our reporting that there is no significant difference.

The important characteristic in this example, however, is the interaction. By examining the graph, we can see that the means for male public school and female private school are not different from each other. However, they are both 20 points higher than male private school and female public school (which are also not different from each other). That is what an interaction is all about. Even though the independent variables by themselves would not produce a difference, the combinations of two or more variables might.

LAYOUT OF THE TWO-BETWEN ANOVA

Table 14.1 shows how two variables (denoted as Variable A and Variable B) would combine together for a two-between ANOVA. There are a few new notations, which are important to note about the one-within ANOVA. In addition, there are some new summary notations with which you should become familiar. These include the following:

Table 14.1: Basic Layout for a Two-Between ANOVA

Independent Variable A	Independent Variable B				
	Level 1	Level 2	Level 3	...	Level K
Level 1	$X_{1,1,1}$ $X_{2,1,1}$ $X_{3,1,1}$ $X_{4,1,1}$ . . . $X_{n,1,1}$	$X_{1,1,2}$ $X_{2,1,2}$ $X_{3,1,2}$ $X_{4,1,2}$ . . . $X_{n,1,2}$	$X_{1,1,3}$ $X_{2,1,3}$ $X_{3,1,3}$ $X_{4,1,3}$ . . . $X_{n,1,3}$	... ...	$X_{1,1,K}$ $X_{2,1,K}$ $X_{3,1,K}$ $X_{4,1,K}$ . . . $X_{n,1,K}$
Level 2	$X_{1,2,1}$ $X_{2,2,1}$ $X_{3,2,1}$ $X_{4,2,1}$ . . . $X_{n,2,1}$	$X_{1,2,2}$ $X_{2,2,2}$ $X_{3,2,2}$ $X_{4,2,2}$ . . . $X_{n,2,2}$	$X_{1,2,3}$ $X_{2,2,3}$ $X_{3,2,3}$ $X_{4,2,3}$ . . . $X_{n,2,3}$	... ...	$X_{1,2,K}$ $X_{2,2,K}$ $X_{3,2,K}$ $X_{4,2,K}$ . . . $X_{n,2,K}$
Level 3	$X_{1,3,1}$ $X_{2,3,1}$ $X_{3,3,1}$ $X_{4,3,1}$ . . . $X_{n,3,1}$	$X_{1,3,2}$ $X_{2,3,2}$ $X_{3,3,2}$ $X_{4,3,2}$ . . . $X_{n,3,2}$	$X_{1,3,3}$ $X_{2,3,3}$ $X_{3,3,3}$ $X_{4,3,3}$ . . . $X_{n,3,3}$	... ...	$X_{1,3,K}$ $X_{2,3,K}$ $X_{3,3,K}$ $X_{4,3,K}$ . . . $X_{n,3,K}$
...	. . . .	. . . .	. . . .	... ...	. . . .
Level J	$X_{1,J,1}$ $X_{2,J,1}$ $X_{3,J,1}$ $X_{4,J,1}$ . . . $X_{n,J,1}$	$X_{1,J,2}$ $X_{2,J,2}$ $X_{3,J,2}$ $X_{4,J,2}$ . . . $X_{n,J,2}$	$X_{1,J,3}$ $X_{2,J,3}$ $X_{3,J,3}$ $X_{4,J,3}$ . . . $X_{n,J,3}$	... ...	$X_{1,J,K}$ $X_{2,J,K}$ $X_{3,J,K}$ $X_{4,J,K}$ . . . $X_{n,J,K}$

X_{ijk} = A data value for the ith participant under the jth level of the first independent variable (A) and the kth level of the second independent variable (B).

J = Number of levels of the first independent variable (A).

K = Number of levels of the second independent variable (B).

$M... $ = Grand mean = $\dfrac{\sum\sum\sum X_{ijk}}{JKn}$

M_J = Mean of the J variable = $\dfrac{\sum\sum X_{ijk}}{Kn}$

M_K = Mean of the K variable = $\dfrac{\sum\sum X_{ijk}}{Jn}$

M_{jk} = Cell mean, comprising one level of the first independent variable and one level of the second independent variable = $\dfrac{\sum X_{ijk}}{n}$

PARTITIONING THE SUMS OF SQUARES

The total sum of squares (SS_{TOTAL}) is calculated as you would in a one-between ANOVA. It represents the sum of the squared deviations of each score from the grand mean.

$$SS_{TOTAL} = \sum\sum\sum(X - M...)^2$$

In the one-between ANOVA there was only one between-groups sum of squares. However, in the two-between ANOVA there will be two between-groups sum of squares: sum of squares for the first independent variable (A) and sum of squares for the second independent variable (B). So, instead of a $SS_{BETWEEN}$ term, we will have terms for SS_A and SS_B. Using the letters A and B is generic...when we actually work on an example problem, we substitute the names of the variables of interest in place of the letters. These two calculations are shown below.

$$SS_A = Kn\sum (M_J - M...)^2$$

$$SS_B = Jn\sum (M_K - M...)^2$$

The error term is based on the sum of squares within groups, and is calculated as follows:

$$SS_{WITHIN} = \sum\sum\sum (X_{ijk} - M_{JK})^2$$

$$or \qquad = JK(n-1)\sigma^2_{POOLED}$$

σ^2_{POOLED} is known as the pooled variance, which is the average of the within-group variance estimates.

The final sum of squares is for the interaction, which measures additional variability that can not be explained by the two main independent variables. It is calculated as follows:

$$SS_{INTERACTION} = n \sum\sum (M_{JK} - M_J - M_K + M...)^2$$

MEAN SQUARES AND DEGREES OF FREEDOM

Mean squares are calculated by dividing the sum of squares by the degrees of freedom for each term.

$$MS_{TOTAL} \qquad = \frac{SS_{TOTAL}}{df_{TOTAL}}$$

$$= \frac{\sum\sum\sum(X - M...)^2}{(JKn)-1}$$

$$MS_A \qquad = \frac{SS_A}{df_A}$$

$$= \frac{Kn\sum (M_J - M...)^2}{(J-1)}$$

MS_B $= \dfrac{SS_B}{df_B}$

$= \dfrac{Jn\sum (M_K - M...)^2}{(K-1)}$

$MS_{INTERACTION}$ $= \dfrac{SS_{INTERACTION}}{df_{INTERACTION}}$

$= \dfrac{n \sum\sum (M_{JK} - M_J - M_K + M...)^2}{(J-1)(K-1)}$

MS_{WITHIN} $= \dfrac{SS_{WITHIN}}{df_{WITHIN}}$

$= \dfrac{\sum\sum\sum (X_{ijk} - M_{JK})^2}{[JK(n-1)]}$

AN EXPERIMENTAL EXAMPLE

An interesting phenomenon in the behavioral sciences is a condition known as state dependent learning. Participants typically learn some material and are then given a test to assess their recall of the material. The recall situation is either similar to the situation in which they learned the material, or it is different. The question of interest is "Do we perform better on an assessment when the assessment situation is the same as the situation in which we learned the material?"

Below are some data from a state dependent learning experiment. Participants learned the material either under the influence of alcohol (drunk) or not under the influence of alcohol (sober). They then took the assessment test either under the influence of alcohol (drunk) or not under the influence of alcohol (sober). Thus, we have four groups, each with 8 participants, in a two-between design. The dependent variable is their performance on the assessment test, with higher scores indicating greater performance. The raw data are listed in Table 14.2, and the means and cell variances are listed in Table 14.3.

Table 14.2: Raw Data for State Dependent Learning Example.

LEARNING   RECALL SITUATION
SITUATION

	Drunk	Sober
Drunk	25	18
	27	15
	26	17
	29	20
	24	14
	29	16
	28	21
	25	18
Sober	17	28
	14	32
	18	34
	19	31
	18	29
	22	33
	21	32
	16	30

Table 14.3: Means and Cell Variances for State Dependent Learning Example.

LEARNING   RECALL SITUATION
SITUATION

	Drunk	**Sober**	ROW MEANS
Drunk	M=26.63 σ^2=3.70	M=17.38 σ^2=5.70	M=22.00
Sober	M=18.13 σ^2=6.70	M=31.13 σ^2=4.12	M=24.63
COLUMN MEANS	M=22.38	M=24.25	M...=23.31

$$\sigma^2_{POOLED} = 5.06$$

The total sum of squares (SS_{TOTAL}) is calculated as in the one-between ANOVA. It represents the sum of the squared deviations of each score from the grand mean.

$$SS_{TOTAL} = \sum\sum\sum(X - M...)^2$$

$$\begin{aligned}
= &(25 - 23.31)^2 + (27 - 23.31)^2 + (26 - 23.31)^2 + \\
&(29 - 23.31)^2 + (24 - 23.31)^2 + (29 - 23.31)^2 + \\
&(28 - 23.31)^2 + (25 - 23.31)^2 + (18 - 23.31)^2 + \\
&(15 - 23.31)^2 + (17 - 23.31)^2 + (20 - 23.31)^2 + \\
&(14 - 23.31)^2 + (16 - 23.31)^2 + (21 - 23.31)^2 + \\
&(18 - 23.31)^2 + (17 - 23.31)^2 + (14 - 23.31)^2 + \\
&(18 - 23.31)^2 + (19 - 23.31)^2 + (18 - 23.31)^2 + \\
&(22 - 23.31)^2 + (21 - 23.31)^2 + (16 - 23.31)^2 + \\
&(28 - 23.31)^2 + (32 - 23.31)^2 + (34 - 23.31)^2 + \\
&(31 - 23.31)^2 + (29 - 23.31)^2 + (33 - 23.31)^2 + \\
&(32 - 23.31)^2 + (30 - 23.31)^2
\end{aligned}$$

$$= 2.86 + 13.62 + 7.24 + 32.38 + .48 + 32.38 + 22.0 +$$
$$2.86 + 28.20 + 69.06 + 39.82 + 10.96 + 86.68 + 53.44$$
$$+ 5.34 + 28.2 + 39.82 + 86.68 + 28.2 + 18.58 + 28.2 +$$
$$1.72 + 5.34 + 53.44 + 22.0 + 75.52 + 114.28 + 59.14 +$$
$$32.38 + 93.9 + 75.52 + 44.76$$

$$= 1215$$

$SS_{LEARNING}$ $= Kn\sum (M_J - M...)^2$

$= (2)(8)[(22 - 23.31)^2 + (24.63 - 23.31)^2]$

$= (2)(8)[1.72 + 1.74]$

$= (2)(8)(3.46)$

$= 55.36$

SS_{RECALL} $= Jn\sum (M_K - M...)^2$

$= (2)(8)[(22.38 - 23.31)^2 + (24.25 - 23.31)^2]$

$= (2)(8)[.86 + .88]$

$= (2)(8)(1.74)$

$= 27.84$

SS_{WITHIN} $= JK(n-1)\sigma^2_{pooled}$

$= (2)(2)(7)(5.06)$

$= 141.68$

$$SS_{INTERACTION} = SS_{TOTAL} - SS_{LEARNING} - SS_{RECALL} - SSWITHIN$$

$$= 1215 - 55.36 - 27.84 - 141.68$$

$$= 990.12$$

The beginnings of the ANOVA table are shown below:

ANOVA Table for State Dependent Learning Example

Source	SS	df	MS	F	p
Total	1215				
Learning	55.36				
Recall	27.84				
Interaction	990.12				
Within	141.68				

The next step is the addition of the degrees of freedom, remembering that...

$$df_{TOTAL} = (JKn)-1$$
$$df_{LEARNING} = J-1$$
$$df_{RECALL} = K-1$$
$$df_{INTERACTION} = (J-1)(K-1)$$
$$df_{WITHIN} = JK(n-1)$$

ANOVA Table for State Dependent Learning Example

Source	SS	df	MS	F	p
Total	1215	31			
Learning	55.36	1			
Recall	27.84	1			
Interaction	990.12	1			
Within	141.68	28			

Next we calculate the mean square by dividing the sum of squares by the degrees of freedom. We are interested only in the two main independent variables, the interaction, and the within term.

ANOVA Table for State Dependent Learning Example

Source	SS	df	MS	F	p
Total	1215	31			
Learning	55.36	1	55.36		
Recall	27.84	1	27.84		
Interaction	990.12	1	990.12		
Within	141.68	28	5.06		

Our final step is obtain a critical value for F and to calculate our actual F score. To obtain the critical value, we again use the F distribution (located in the back of your manual). The degrees of freedom for the numerator are the same as the degrees of freedom for our learning variable, recall variable, and interaction. The degrees of freedom for the denominator are the same as the degrees of freedom for our within source of variability.

In this example, we go across the top line until we find 1 and along the left side until we find 28. With $\alpha=.05$ (lightface type) and 1 and 28 degrees of freedom, we see that the critical value of F is 4.20, so $F'(1,28) = 4.20$.

Notice the difference in the two-between ANOVA, in that we now have 3 different F scores for comparison--one for each of the two main effects (Learning and Recall) and one for the interaction effect. If the score for our F tests is 4.20 or greater, we will say there is a significant effect, and reject the null hypothesis. If the score for our F tests is less than 4.20, we will say there is not a significant effect, and accept the null hypothesis.

The actual F value is calculated by dividing the mean square for the main effect variables and the interaction variable by MS_{WITHIN}. In our example, we get $55.36 \div 5.06 = 10.94$ for Learning, $27.84 \div 5.06 = 5.50$ for Recall, and $990.12 \div 5.06 = 195.68$ for the interaction. Since all of these values are greater than our critical value of 4.20, we reject the null hypothesis for Learning, Recall and the Interaction.

The final ANOVA table would take on the following appearance:

ANOVA Table for State Dependent Learning Example

Source	SS	df	MS	F	p
Total	1215	31			
Learning	55.36	1	55.36	10.94	<.05
Recall	27.84	1	27.84	5.50	<.05
Interaction	990.12	1	990.12	195.68	<.05
Within	141.68	28	5.06		

What conclusions can we make? In a two between ANOVA, we need to make three separate sets of conclusions: two for the main effect variables (A and B), and one for the interaction variable. We will begin with the main effects.

We found a significant difference for Learning. Looking back at our table of means, we see that the mean for learning drunk (M=22.0) is less than the mean for learning sober (M=24.63). So our first conclusion is that people perform better on the assessment when they learn sober.

We also found a significant difference for Recall. Again checking the table of means, we see that the mean for recall drunk (M=22.38) is less than the mean for recall sober (M=24.25). Our second conclusion is that people perform better on the assessment when they recall sober.

Finally, we found a significant effect for the interaction. The best way to evaluate interaction effects is through plotting the data, which is done in Figure 14.2.

Figure 14.2: Bar Graph of State Dependent Learning Example Means

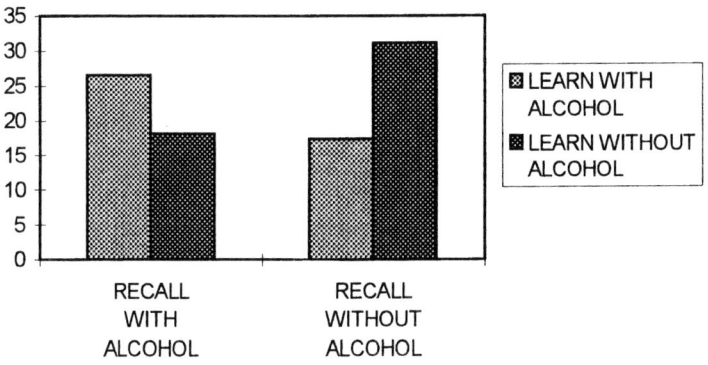

From the graph we can see that people perform better when they both learn and recall in the same state (either learn and recall with alcohol or learn and recall without alcohol).

When dealing with interaction effects, it is sometimes easier to interpret the interactions when the data are plotted in a line graph, as in Figure 14.3.

Figure 14.3: **Line Graph of State Dependent Learning Means**

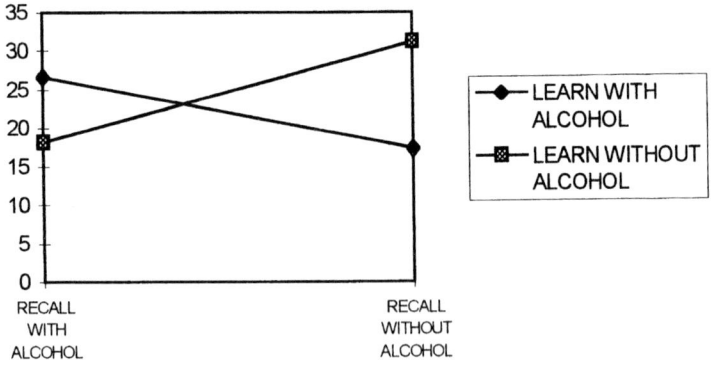

Interaction effects are noted when the lines of our graph either cross (as they do above), or when they deviate from being parallel. However, we can not just look at a graph to determine whether or not there is a statistically significant interaction. The first step in assessing the existence of an interaction is through the ANOVA.

ACTIVITY 14.1: A researcher was interested in examining maze running ability in rats. Two variables of interest were the type of food the rats had been raised on and previous experience in other mazes.

Below are some data from an experiment combining these two variables. Rats ate either rat chow or a special blend diet. They also either had previous maze running experience, or no maze running experience. The dependent variable was how long it took the rat (in seconds) to find its way to the end of the maze, and thus receive a reward.

FOOD TYPE	MAZE EXPERIENCE	
	None	**Experienced**
Rat Chow	10	11
	15	13
	12	12
	18	16
	17	14
Special Blend	12	5
	18	7
	14	10
	14	8
	10	6

MEANS AND CELL VARIANCES

FOOD TYPE	MAZE EXPERIENCE		
	None	**Experienced**	ROW MEANS
Rat Chow	M=14.4	M=13.2	M=13.8
	σ^2=11.3	σ^2=13.2	
Special Blend	M=13.6	M=7.2	M=10.4
	σ^2=8.80	σ^2=3.7	
COLUMN MEANS	M=14.0	M=10.2	M...=12.1

$\sigma^2_{POOLED} = 6.88$

$$SS_{TOTAL} = \sum\sum\sum(X - M...)^2$$

$$SS_{FOOD} = Kn\sum (M_J - M...)^2$$

$$SS_{EXPERIENCE} = Jn\sum (M_K - M...)^2$$

$$SS_{WITHIN} = JK(n-1)\sigma^2_{POOLED}$$

$$SS_{INTERACTION} = SS_{TOTAL} - SS_{FOOD} - SS_{EXPERIENCE} - SS_{WITHIN}$$

Source	SS	df	MS	F	p
Total					
Food					
Experience					
Interaction					
Within					

What conclusions can you make based on the outcome of this study? Be sure to plot the means so that you can interpret the interaction effect.

ADDITIONAL ACTIVITIES

1. Two kinds of food-stuffs (beer and coffee) were tested for taste preference in two different types of containers (a beer bottle and a coffee mug).   Higher scores indicate greater beverage preference. The data are as follows:

	BEER	COFFEE
BOTTLE	6	4
	5	4
	6	5
	4	3
	6	5
	7	4
MUG	3	5
	4	7
	4	6
	2	7
	3	6
	4	7

a. Is there a significant main effect for food-stuff type? If so, what?

b. Is there a significant main effect for container type? If so, what?

c. Is there a significant interaction? If so, describe the interaction.

d. Plot the data, with food-stuff on the X axis, and make separate lines for each container.

2. Two kinds of cars (economy and sports) were tested for durability under two types of rough terrain (snow and mud). The testing data are listed below. Higher scores indicate greater durability.

	ECONOMY	SPORTS
SNOW	3	9
	1	8
	2	9
MUD	10	12
	9	10
	8	9

a.   Is there a significant main effect for car type? If so, what?

b.   Is there a significant main effect for terrain type? If so, what?

c.   Is there a significant interaction? If so, describe the interaction.

d.   Plot the data, with car type on the X axis, and make separate lines for each terrain.

MODULE XV: CHI-SQUARE TEST OF INDEPENDENCE

THE LOGIC OF THE CHI SQUARE (X^2)

Up to this point, we have addressed statistical tests where we are interested in determining whether a difference exists between or among means. What happens when we do not have means to compare? That is where the Chi Square (X^2) test comes into play. The Chi Square is a sampling distribution, much like the t or F distributions, although it gives probabilities associated with frequencies of occurrence, rather than means. What distinguishes this test from the others is the type of data with which we will be dealing, which are frequency counts.

Often times it is important to determine whether or not there is a difference in the frequency of the occurrence of some event. For example, is there a statistical difference in the number of students enrolled in the various majors offered by a university? Or, is there a statistical difference in the proportion of cancer cases noted in one particular area over another? Or, are the proportions of males vs. females different in their attitudes toward gun control? In essence, we are asking whether one event is independent of another event. We will further examine this last question, and determine whether or not the proportions of males vs. females differ in their attitudes toward gun control.

AN EXPERIMENTAL EXAMPLE

From a sample of 1,000 persons we obtain the following crosstabulation of sex and gun control attitude. The scores represent the number of males and females who either support or oppose a new gun control legislation.

	SUPPORT	OPPOSE	Σ
MALES	460	180	640
FEMALES	120	240	360
Σ	580	420	1000

OBSERVED AND EXPECTED FREQUENCIES

The above crosstabulation contains what are called observed frequencies (denoted as f_o), which are the actual frequencies from our study. In order to determine whether the variables of sex and gun control attitude are independent, we need to compare the observed frequencies with expected frequencies (denoted as f_e). Expected frequencies are the frequencies we would expect based on the normal laws of probability

The formula for expected frequencies is as follows:

$$f_e = \frac{(R * C)}{N}$$

where R is the row marginal frequency and C is the column marginal frequency.

Calculating the expected frequencies for this example would give us the following scores:

MALES-SUPPORT: $\frac{(640 * 580)}{1000}$ = 371.2

MALES-OPPOSE: $\frac{(640 * 420)}{1000}$ = 268.8

FEMALES-SUPPORT: $\dfrac{(360 * 580)}{1000}$ $= 208.8$

FEMALES-OPPOSE: $\dfrac{(360 * 420)}{1000}$ $= 151.2$

The following chart integrates both the observed and the expected values.

f_o f_e	SUPPORT	OPPOSE
MALES	460 371.2	180 268.8
FEMALES	120 208.8	240 151.2

TESTING THE NULL HYPOTHESIS OF INDEPENDENCE

We take as our test statistic:

$$X^2 = \dfrac{\Sigma(f_o - f_e)^2}{f_e}$$

with df $= (R - 1)(C - 1)$

In our example, the X^2 is calculated as follows:

X^2 $= \dfrac{(460 - 371.2)^2}{371.2} + \dfrac{(180 - 268.8)^2}{268.8} + \dfrac{(120 - 208.8)^2}{208.8} +$

$\dfrac{(240 - 151.2)^2}{151.2}$

$= 21.24 + 29.34 + 37.77 + 52.15$

$= 140.5$

A significant X^2 indicates that the proportions you find for the different levels of one variable are dependent upon the different levels of the other variable.

To be significant, the X^2 obtained from the data must be equal to or larger than the critical value of X^2. The critical value can be obtained from the X^2 table (located in the back of your manual). With $\alpha=.05$ and degrees of freedom (R-1)(C-1), or (1)(1) = 1, the critical value is 3.84 Therefore, $X^2(1)'=3.84$. Since our actual X^2 value of 140.5 is greater than the critical value of 3.84, we can say that the test is statistically significant and reject the null hypothesis that our variables are independent. Therefore, we can conclude that the attitudes about gun control are influenced by the sex of the person, with more males supporting and fewer females supporting the gun control legislation that you would have expected based on the laws of normal probability.

ACTIVITY 15.1: A psychology student was interested in whether or not gender and major were independent events. She obtained information on 500 students and composed the following table of observed values.

	Natural Sciences	Behavioral Sciences	Liberal Arts	Business	Σ
Male	85	55	33	40	213
Female	70	88	84	45	287
Σ	155	143	117	85	500

Perform the appropriate test to determine whether sex of the participant and college major are independent events.

ADDITIONAL ACTIVITIES

1. A social psychologist hypothesized that a factor in juvenile delinquency was the presence or absence of a strong father-figure in the home. She examined the folders of 50 male inmates in the federal reformatory and found that 25 of these young men grew up with a strong father-figure in the home. She also examined the records of 100 randomly selected male college students and found that 75 of them had strong father-figures in their boyhood homes. Perform the appropriate analysis to test the psychologist's hypothesis, and comment on the results.

2. An market research analyst was interested in whether or not there was an effect of color on the type of car purchased. She obtained a random sample of 250 car sales and recorded the car colors as follows: Red: 45, Green: 37, Blue: 48, White: 64, Black: 56. Perform the appropriate analysis to test the hypothesis, and comment on the results.

MODULE XVI: NONPARAMETRIC PROCEDURES

USE OF NONPARAMETRIC PROCEDURES

Until this point, we have primarily discussed what are known as parametric tests, such as the t-test, ANOVA, Tukey HSD, etc. Parametric tests rely on parameters such as the mean, variance, standard deviation, etc. in order to describe or provide inference concerning a data set. There are times, however, when it is more appropriate to use non-parametric tests. Non-parametric tests are used when the dependent variable consists of nominal or ordinal data, rather than interval or ratio data in which there are continuous scores.

The following sections will provide an introduction to some of the more widely used nonparametric procedures. Be sure to note the relationship between these nonparametric tests and their parametric counterparts.

MANN-WHITNEY U TEST OR WILCOXON RANK-SUM TEST

The Mann-Whitney U test (also known as the Wilcoxon rank sum test) is the best alternative to the t-test for two independent means. It is not necessary that the two samples be of equal size. The test is based on ranking all scores, regardless of group, and the resulting statistic is known as a U.

AN EXPERIMENTAL EXAMPLE

A researcher believed that class rank in college may be related to having gone to a private high school vs. a public high school. The following data relate to the top 10 students (academically speaking) in a particular college, and whether they went to a public or private high school.

Rank	High School
1	Private
2	Public
3	Private
4	Private
5	Public
6	Private
7	Public
8	Public
9	Private
10	Public

We begin by summing the ranks for each of the high schools

$Rank_1$ (Private) $= 1 + 3 + 4 + 6 + 9 = 23$

$Rank_2$ (Public) $= 2 + 5 + 7 + 8 + 10 = 32$

Next, we calculate U for each group, where

$$U_i = N_1N_2 + [N_i(N_i + 1)/2] - Rank_i$$

and i=the particular rank of interest.

For the private school,

$$U_{PRIVATE} = (5)(5) + [5(5+1)/2] - 23$$

$$= 25 + 15 - 23$$

$$= 17$$

For the public school,

$$U_{PUBLIC} = (5)(5) + [5(5+1)/2] - 32$$

$$= 25 + 15 - 32$$

$$= 8$$

We will make a comparison based on the lowest value of U. In this case, the lowest U value is 8. We refer this result to the table of critical values of the Mann-Whitney U test (in the back of your manual). We see that with $N_1 = 5$ and $N_2 = 5$ for a two-tailed test with $\alpha = .02$, the critical value of U is 1. Any value of U that is equal to or <u>less than</u> the critical value leads to the decision to reject the null hypothesis. Note carefully that this procedure for accepting or rejecting the null hypothesis is different from what we have previously learned.

In our example, the U value of 8 is not less than the critical value of 1, so we do not reject the null hypothesis. Therefore, we conclude that either going to a private or public school has no influence on class rank.

ACTIVITY 16.1: After spending some time at your local greyhound dog racing track, you have heard that performance of the greyhounds is influenced by whether or not the dogs "relieve" themselves right before the race. To test this hypothesis, the following data indicate the place in which the dog finished, and whether or not the dog "relieved" himself before the race.

Place	"Relieved" himself
1	Yes
2	Yes
3	Yes
4	Yes
5	No
6	Yes
7	No
8	No

Perform the appropriate analysis to determine whether or not there is a statistically significant effect.

WILCOXON MATCHED-PAIRS SIGNED-RANKS T-TEST

This procedure is the best alternative to the t-test for two related means. It will allow you to determine whether there has been a change from one time to the next (ie., pre-post).

AN EXPERIMENTAL EXAMPLE

We have asked people to rate their level of self-confidence on a scale from 0-10, both before a self-esteem workshop and after. Their scores are below. Also below are their difference scores (before vs. after). Note that it does not matter in which manner you perform the difference score (that is to say, which value you subtract first), as long as you are consistent throughout.

There are, however, some cautionary notes concerning the case where the difference score = 0. When the difference score = 0, the pair of scores is dropped from the analysis and n is reduced by 1. When two of the difference scores = 0, each is given a rank, with one being assigned a positive sign and the other being assigned a negative sign. If three of the difference scores = 0, one is dropped, n is reduced by 1, and the remaining two are given oppositely signed ranks. And so on.

Participant	Before	After	Difference Score
1	7	8	-1
2	6	4	2
3	8	9	-1
4	10	10	0
5	8	7	1
6	9	9	0
7	7	6	1
8	5	7	-2
9	9	6	3

We first rank the absolute value of the difference scores from low to high and then sign the ranks.

Difference Score	0	0	1	1	1	1	2	2	3
Rank	1.5	-1.5	4.5	4.5	4.5	4.5	7.5	7.5	9
Signed Rank	1.5	-1.5	-4.5	-4.5	4.5	4.5	-7.5	7.5	-9

Note that when there are ties, we need to make an average rating for all the scores that receive that ranking. For example, since there are two 0's, they would take the rankings of 1 and 2; the average of 1 and 2 is 1.5, so they each receive a 1.5. Since there are four 1's, they would take the rankings of 3, 4, 5, and 6; the average of 3, 4, 5, and 6 is 4.5, so they each receive a 4.5

We base the statistical analysis on the sign that occurs with the least frequently. In our example, that is the positive sign.

The sum of the absolute values of the positive ranks is computed and given the notation T. Do not confuse this T with the t-test you have learned earlier.

For this example,

$$T \quad = 1.5 + 4.5 + 4.5 + 7.5$$
$$= 18$$

Referring to a table of critical values (in the back of your textbook), the critical value of T for 9 pairs of values, 2-tailed, for $\alpha = .05$, is 5. T must be equal to or less than the critical value. We therefore fail to reject the null hypothesis, since our T value of 18 is greater than the critical value of 5. We can therefore conclude that there is no significant difference in the pre vs. post self-esteem scores.

ACTIVITY 16.2: A physiological psychologist wanted to check pupil dilation before and after the presentation of a new drug. Dilation measurements, in millimeters, are below. Also below are the difference scores (pre vs. post drug).

Participant	Before	After	Difference Score
1	4	8	4
2	6	7	1
3	5	7	2
4	7	8	1
5	4	6	2
6	5	7	2
7	4	8	4
8	8	8	0
9	6	7	1
10	4	6	2

Perform the appropriate analysis to determine if there is a significant change in pupil dilation pre vs. post.

WILCOXON-WILCOX   MULTIPLE COMPARISONS TEST

The Wilcoxon-Wilcox Multiple Comparisons Test allows you to compare three or more independent groups. Thus, it is the nonparametric equivalent of the Between Groups ANOVA.

It is important to note that the number of participants in each group, with group being designated by the letter K, must be equal. In the situation where you have different numbers of observations in each group, one approach would be to randomly eliminate members of a group until all groups have an equal number of observations.

AN EXPERIMENTAL EXAMPLE

Suppose that we were interested in how well four different cars performed with the same type of gasoline. We filled each car with gasoline five times, each time allowing for normal driving to determine the miles per gallon. These data are presented below:

Car A	Car B	Car C	Car D
25.5	28.4	30.2	34.5
23.2	26.7	29.6	33.8
24.4	25.3	27.8	37.5
21.9	27.2	31.0	35.2
22.0	27.9	29.9	33.7

The first step is to rank all values, regardless of the group, and compute a sum of ranks for each group.

Car A	Car A Rank	Car B	Car B Rank	Car C	Car C Rank	Car D	Car D Rank
25.5	6	28.4	11	30.2	14	34.5	18
23.2	3	26.7	7	29.6	12	33.8	17
24.4	4	25.3	5	27.8	9	37.5	20
21.9	1	27.2	8	31.0	15	35.2	19
22.0	2	27.9	10	29.9	13	33.7	16

Σ 16   41   63   90

The next step is to make comparisons between pairs. With four groups, there are six possible pairs of comparisons. To do this, we take the difference between the ranks, and obtain the following values:

Car A vs. Car B: 16 - 41 = 25

Car A vs. Car C: 16 - 63 = 47

Car A vs. Car D: 16 - 90 = 74

Car B vs. Car C: 41 - 63 = 22

Car B vs. Car D: 41 - 90 = 49

Car C vs. Car D: 63 - 90 = 27

Since we are interested only in the absolute difference between the pairs, that is the number that has been presented above.

The final step is to determine which pairs are significantly different. The critical value table (in the back of your textbook) shows values for different levels of K (that is, different numbers of independent groups). In our example, K=4, so with N=5 and α=.05, the critical value is 48.1. To be significant, our difference between the groups must be 48.1 or greater. We can see that we have two significant differences in the car data: Car A vs. Car D (at 74) and Car B vs. Car D (at 49). Therefore, we can conclude that gas millage ranking is higher in Car D than either Car A or Car B.

ACTIVITY 16.3: Suppose that we were interested in how well three different batteries performed in the same type of electronic video game. We placed the same type of batter in the game 8 times, each time allowing for normal video game playing use. These data are presented below, which indicate the number of hours of play time.

Battery A	Battery B	Battery C
156	160	184
184	174	197
165	188	179
145	162	188
165	170	186
144	159	195
138	188	197
150	173	193

Rank all values, regardless of the group, and compute a sum of ranks for each group. Then, perform the statistical test to determine if there is a significant difference among any of the batteries. What can you conclude from the analyses?

SPEARMAN'S RHO

A widely used rank order correlation coefficient, designed for assessing the relationship between two variables, is Spearman's rho. We will abbreviate it as r_s.

The first step in calculating r_s is to convert to ranks all of the scores in each variable, and then calculate the sum of the squared rank differences.

AN EXPERIMENTAL EXAMPLE

We are interested in whether or not there is a relationship between age and yearly salary of the top executives in America. We gather the following data, which indicate rank in age from youngest to oldest, and rank in salary from lowest to highest.

Rank in Age (X)	Rank in Salary (Y)	d	d²
1	9	-8	64
2	10	-8	64
3	7	-4	16
4	8	-4	16
5	4	1	1
6	6	0	0
7	3	4	16
8	2	6	36
9	5	4	16
10	1	9	81

$$\Sigma = 310$$

The formula for Spearman's rho is as follows:

$$r_s = 1 - \frac{6\Sigma d^2}{N(N^2 - 1)}$$

In our example,

$$r_s = 1 - \frac{6*310}{10(100 - 1)}$$

$$= 1 - \frac{1860}{990}$$

$$= 1 - 1.88$$

$$= -.88$$

To be significant, the r_s obtained from the data must be equal to or larger than the value shown in the critical value table (in the back of your textbook). In our example, with 10 pairs of scores, a two-tailed test, and α = .05, the critical value is 0.648. Since our value of -.88 (remember, the sign only tells us the direction of the effect, not the magnitude) is greater than the critical value of 0.648, we can reject the null hypothesis and conclude that there is a significant relationship between age and salary. Specifically, as age rank increases, salary rank decreases, thus a greater salary is associated with an older age.

ACTIVITY 16.4: A researcher was interested in the possible relationship between tenure of University professors and the number of publications they have produced. He gathered the following data, which indicate rank in tenure from most recent tenure to longest tenure, and rank in number of publications from least to greatest. These data are presented below.

Rank in Tenure (X)	Rank in Publications (Y)
1	13
2	15
3	14
4	11
5	8
6	12
7	10
8	5
9	9
10	6
11	3
12	7
13	4
14	1
15	2

Perform the appropriate analysis to determine if a relationship exists between tenure and publication ranks.

APPENDIX A:
STATISTICAL TABLES

Areas Under the Normal Curve

A z	B Area between mean and z	C Area beyond z	A z	B Area between mean and z	C Area beyond z	A z	B Area between mean and z	C Area beyond z
0.00	.0000	.5000	0.55	.2088	.2912	1.10	.3643	.1357
0.01	.0040	.4960	0.56	.2123	.2877	1.11	.3665	.1335
0.02	.0080	.4920	0.57	.2157	.2843	1.12	.3686	.1314
0.03	.0120	.4880	0.58	.2190	.2810	1.13	.3708	.1292
0.04	.0160	.4840	0.59	.2224	.2776	1.14	.3729	.1271
0.05	.0199	.4801	0.60	.2257	.2743	1.15	.3749	.1251
0.06	.0239	.4761	0.61	.2291	.2709	1.16	.3770	.1230
0.07	.0279	.4721	0.62	.2324	.2676	1.17	.3790	.1210
0.08	.0319	.4681	0.63	.2357	.2643	1.18	.3810	.1190
0.09	.0359	.4641	0.64	.2389	.2611	1.19	.3830	.1170
0.10	.0398	.4602	0.65	.2422	.2578	1.20	.3849	.1151
0.11	.0438	.4562	0.66	.2454	.2546	1.21	.3869	.1131
0.12	.0478	.4522	0.67	.2486	.2514	1.22	.3888	.1112
0.13	.0517	.4483	0.68	.2517	.2483	1.23	.3907	.1093
0.14	.0557	.4443	0.69	.2549	.2451	1.24	.3925	.1075
0.15	.0596	.4404	0.70	.2580	.2420	1.25	.3944	.1056
0.16	.0636	.4364	0.71	.2611	.2389	1.26	.3962	.1038
0.17	.0675	.4325	0.72	.2642	.2358	1.27	.3980	.1020
0.18	.0714	.4286	0.73	.2673	.2327	1.28	.3997	.1003
0.19	.0753	.4247	0.74	.2704	.2296	1.29	.4015	.0985
0.20	.0793	.4207	0.75	.2734	.2266	1.30	.4032	.0968
0.21	.0832	.4168	0.76	.2764	.2236	1.31	.4049	.0951
0.22	.0871	.4129	0.77	.2794	.2206	1.32	.4066	.0934
0.23	.0910	.4090	0.78	.2823	.2177	1.33	.4082	.0918
0.24	.0948	.4052	0.79	.2852	.2148	1.34	.4099	.0901
0.25	.0987	.4013	0.80	.2881	.2119	1.35	.4115	.0885
0.26	.1026	.3974	0.81	.2910	.2090	1.36	.4131	.0869
0.27	.1064	.3936	0.82	.2939	.2061	1.37	.4147	.0853
0.28	.1103	.3897	0.83	.2967	.2033	1.38	.4162	.0838
0.29	.1141	.3859	0.84	.2995	.2005	1.39	.4177	.0823
0.30	.1179	.3821	0.85	.3023	.1977	1.40	.4192	.0808
0.31	.1217	.3783	0.86	.3051	.1949	1.41	.4207	.0793
0.32	.1255	.3745	0.87	.3078	.1922	1.42	.4222	.0778
0.33	.1293	.3707	0.88	.3106	.1894	1.43	.4236	.0764
0.34	.1331	.3669	0.89	.3133	.1867	1.44	.4251	.0749
0.35	.1368	.3632	0.90	.3159	.1841	1.45	.4265	.0735
0.36	.1406	.3594	0.91	.3186	.1814	1.46	.4279	.0721
0.37	.1443	.3557	0.92	.3212	.1788	1.47	.4292	.0708
0.38	.1480	.3520	0.93	.3238	.1762	1.48	.4306	.0694
0.39	.1517	.3483	0.94	.3264	.1736	1.49	.4319	.0681
0.40	.1554	.3446	0.95	.3289	.1711	1.50	.4332	.0668
0.41	.1591	.3409	0.96	.3315	.1685	1.51	.4345	.0655
0.42	.1628	.3372	0.97	.3340	.1660	1.52	.4357	.0643
0.43	.1664	.3336	0.98	.3365	.1635	1.53	.4370	.0630
0.44	.1700	.3300	0.99	.3389	.1611	1.54	.4382	.0618
0.45	.1736	.3264	1.00	.3413	.1587	1.55	.4394	.0606
0.46	.1772	.3228	1.01	.3438	.1562	1.56	.4406	.0594
0.47	.1808	.3192	1.02	.3461	.1539	1.57	.4418	.0582
0.48	.1844	.3156	1.03	.3485	.1515	1.58	.4429	.0571
0.49	.1879	.3121	1.04	.3508	.1492	1.59	.4441	.0559
0.50	.1915	.3085	1.05	.3531	.1469	1.60	.4452	.0548
0.51	.1950	.3050	1.06	.3554	.1446	1.61	.4463	.0537
0.52	.1985	.3015	1.07	.3577	.1423	1.62	.4474	.0526
0.53	.2019	.2981	1.08	.3599	.1401	1.63	.4484	.0516
0.54	.2054	.2946	1.09	.3621	.1379	1.64	.4495	.0505

(*continued*)

(continued)

A z	B Area between mean and z	C Area beyond z	A z	B Area between mean and z	C Area beyond z	A z	B Area between mean and z	C Area beyond z
1.65	.4505	.0495	2.22	.4868	.0132	2.79	.4974	.0026
1.66	.4515	.0485	2.23	.4871	.0129	2.80	.4974	.0026
1.67	.4525	.0475	2.24	.4875	.0125	2.81	.4975	.0025
1.68	.4535	.0465	2.25	.4878	.0122	2.82	.4976	.0024
1.69	.4545	.0455	2.26	.4881	.0119	2.83	.4977	.0023
1.70	.4554	.0446	2.27	.4884	.0116	2.84	.4977	.0023
1.71	.4564	.0436	2.28	.4887	.0113	2.85	.4978	.0022
1.72	.4573	.0427	2.29	.4890	.0110	2.86	.4979	.0021
1.73	.4582	.0418	2.30	.4893	.0107	2.87	.4979	.0021
1.74	.4591	.0409	2.31	.4896	.0104	2.88	.4980	.0020
1.75	.4599	.0401	2.32	.4898	.0102	2.89	.4981	.0019
1.76	.4608	.0392	2.33	.4901	.0099	2.90	.4981	.0019
1.77	.4616	.0384	2.34	.4904	.0096	2.91	.4982	.0018
1.78	.4625	.0375	2.35	.4906	.0094	2.92	.4982	.0018
1.79	.4633	.0367	2.36	.4909	.0091	2.93	.4983	.0017
1.80	.4641	.0359	2.37	.4911	.0089	2.94	.4984	.0016
1.81	.4649	.0351	2.38	.4913	.0087	2.95	.4984	.0016
1.82	.4656	.0344	2.39	.4916	.0084	2.96	.4985	.0015
1.83	.4664	.0336	2.40	.4918	.0082	2.97	.4985	.0015
1.84	.4671	.0329	2.41	.4920	.0080	2.98	.4986	.0014
1.85	.4678	.0322	2.42	.4922	.0078	2.99	.4986	.0014
1.86	.4686	.0314	2.43	.4925	.0075	3.00	.4987	.0013
1.87	.4693	.0307	2.44	.4927	.0073	3.01	.4987	.0013
1.88	.4699	.0301	2.45	.4929	.0071	3.02	.4987	.0013
1.89	.4706	.0294	2.46	.4931	.0069	3.03	.4988	.0012
1.90	.4713	.0287	2.47	.4932	.0068	3.04	.4988	.0012
1.91	.4719	.0281	2.48	.4934	.0066	3.05	.4989	.0011
1.92	.4726	.0274	2.49	.4936	.0064	3.06	.4989	.0011
1.93	.4732	.0268	2.50	.4938	.0062	3.07	.4989	.0011
1.94	.4738	.0262	2.51	.4940	.0060	3.08	.4990	.0010
1.95	.4744	.0256	2.52	.4941	.0059	3.09	.4990	.0010
1.96	.4750	.0250	2.53	.4943	.0057	3.10	.4990	.0010
1.97	.4756	.0244	2.54	.4945	.0055	3.11	.4991	.0009
1.98	.4761	.0239	2.55	.4946	.0054	3.12	.4991	.0009
1.99	.4767	.0233	2.56	.4948	.0052	3.13	.4991	.0009
2.00	.4772	.0228	2.57	.4949	.0051	3.14	.4992	.0008
2.01	.4778	.0222	2.58	.4951	.0049	3.15	.4992	.0008
2.02	.4783	.0217	2.59	.4952	.0048	3.16	.4992	.0008
2.03	.4788	.0212	2.60	.4953	.0047	3.17	.4992	.0008
2.04	.4793	.0207	2.61	.4955	.0045	3.18	.4993	.0007
2.05	.4798	.0202	2.62	.4956	.0044	3.19	.4993	.0007
2.06	.4803	.0197	2.63	.4957	.0043	3.20	.4993	.0007
2.07	.4808	.0192	2.64	.4959	.0041	3.21	.4993	.0007
2.08	.4812	.0188	2.65	.4960	.0040	3.22	.4994	.0006
2.09	.4817	.0183	2.66	.4961	.0039	3.23	.4994	.0006
2.10	.4821	.0179	2.67	.4962	.0038	3.24	.4994	.0006
2.11	.4826	.0174	2.68	.4963	.0037	3.25	.4994	.0006
2.12	.4830	.0170	2.69	.4964	.0036	3.30	.4995	.0005
2.13	.4834	.0166	2.70	.4965	.0035	3.35	.4996	.0004
2.14	.4838	.0162	2.71	.4966	.0034	3.40	.4997	.0003
2.15	.4842	.0158	2.72	.4967	.0033	3.45	.4997	.0003
2.16	.4846	.0154	2.73	.4968	.0032	3.50	.4998	.0002
2.17	.4850	.0150	2.74	.4969	.0031	3.60	.4998	.0002
2.18	.4854	.0146	2.75	.4970	.0030	3.70	.4999	.0001
2.19	.4857	.0143	2.76	.4971	.0029	3.80	.4999	.0001
2.20	.4861	.0139	2.77	.4972	.0028	3.90	.49995	.00005
2.21	.4864	.0136	2.78	.4973	.0027	4.00	.49997	.00003

Correlation Table

df	.10	.05	.02	.01	.001
	α Levels (two-tailed test)				
	α Levels (one-tailed test)				
(df = N − 2)	.05	.025	.01	.005	.0005
1	.98769	.99692	.999507	.999877	.9999988
2	.90000	.95000	.98000	.990000	.99900
3	.8054	.8783	.93433	.95873	.99116
4	.7293	.8114	.8822	.91720	.97406
5	.6694	.7545	.8329	.8745	.95074
6	.6215	.7067	.7887	.8343	.92493
7	.5822	.6664	.7498	.7977	.8982
8	.5494	.6319	.7155	.7646	.8721
9	.5214	.6021	.6851	.7348	.8371
10	.4973	.5760	.6581	.7079	.8233
11	.4762	.5529	.6339	.6835	.8010
12	.4575	.5324	.6120	.6614	.7800
13	.4409	.5139	.5923	.6411	.7603
14	.4259	.4973	.5742	.6226	.7420
15	.4124	.4821	.5577	.6055	.7246
16	.4000	.4683	.5425	.5897	.7084
17	.3887	.4555	.5285	.5751	.6932
18	.3783	.4438	.5155	.5614	.6787
19	.3687	.4329	.5034	.5487	.6652
20	.3598	.4227	.4921	.5368	.6524
25	.3233	.3809	.4451	.4869	.5974
30	.2960	.3494	.4093	.4487	.5541
35	.2746	.3246	.3810	.4182	.5189
40	.2573	.3044	.3578	.3932	.4896
45	.2428	.2875	.3384	.3721	.4648
50	.2306	.2732	.3218	.3541	.4433
60	.2108	.2500	.2948	.3248	.4078
70	.1954	.2319	.2737	.3017	.3799
80	.1829	.2172	.2565	.2830	.3568
90	.1726	.2050	.2422	.2673	.3375
100	.1638	.1946	.2301	.2540	.3211

* To be significant the r obtained from the data must be equal to or **larger than** the value shown in the table.

SOURCE: From Table VII in R. A. Fisher and F. Yates, *Statistical Tables for Biological, Agricultural, and Medical Research*, Sixth Edition, published by Addison Wesley Longman Ltd., (1963).

t-test Table

df	Confidence interval percents (two-tailed)					
	80%	90%	95%	98%	99%	99.9%
	α level for two-tailed test					
	.20	.10	.05	.02	.01	.001
	α level for one-tailed test					
	.10	.05	.025	.01	.005	.0005
1	3.078	6.314	12.706	31.821	63.657	636.619
2	1.886	2.920	4.303	6.965	9.925	31.598
3	1.638	2.353	3,182	4.541	5.841	12.924
4	1.533	2.132	2.776	3.747	4.604	8.610
5	1.476	2.015	2.571	3.365	4.032	6.869
6	1.440	1.943	2.447	3.143	3.707	5.959
7	1.415	1.895	2.365	2.998	3.499	5.408
8	1.397	1.860	2.306	2.896	3.355	5.041
9	1.383	1.833	2.262	2.821	3.250	4.781
10	1.372	1.812	2.228	2.764	3.169	4.587
11	1.363	1.796	2.201	2.718	3.106	4.437
12	1.356	1.782	2.179	2.681	3.055	4.318
13	1.350	1.771	2.160	2.650	3.012	4.221
14	1.345	1.761	2.145	2.624	2.977	4.140
15	1.341	1.753	2.131	2.602	2.947	4.073
16	1.337	1.746	2.120	2.583	2.921	4.015
17	1.333	1.740	2.110	2.567	2.898	3.965
18	1.330	1.734	2.101	2.552	2.878	3.922
19	1.328	1.729	2.093	2.539	2.861	3.883
20	1.325	1.725	2.086	2.528	2.845	3.850
21	1.323	1.721	2.080	2.518	2.831	3.819
22	1.321	1.717	2.074	2.508	2.819	3.792
23	1.319	1.714	2.069	2.500	2.807	3.767
24	1.318	1.711	2.064	2.492	2.797	3.745
25	1.316	1.708	2.060	2.485	2.787	3.725
26	1.315	1.706	2.056	2.479	2.779	3.707
27	1.314	1.703	2.052	2.474	2.771	3.690
28	1.313	1.701	2.048	2.467	2.763	3.674
29	1.311	1.699	2.045	2.462	2.756	3.659
30	1.310	1.697	2.042	2.457	2.750	3.646
40	1.303	1.684	2.021	2.423	2.704	3.551
60	1.296	1.671	2.000	2.390	2.660	3.460
120	1.289	1.658	1.980	2.358	2.617	3.373
∞	1.282	1.645	1.960	2.326	2.576	3.291

* To be significant the t obtained from the data must be equal to or **larger than** the value shown in the table.

SOURCE: From Table III in R. A. Fisher and F. Yates, *Statistical Tables for Biological, Agricultural, and Medical Research*, Sixth Edition, published by Addison Wesley Longman Ltd., (1974).

F-test Table

α levels of .05 (lightface) and .01 (boldface) for the distribution of F

Degrees of freedom (for the numerator of F ratio)

Degrees of freedom (for the denominator of F ratio)

	1	2	3	4	5	6	7	8	9	10	11	12	14	16	20	24	30	40	50	75	100	200	500	∞
1	161 / 4,052	200 / 4,999	216 / 5,403	225 / 5,625	230 / 5,764	234 / 5,859	237 / 5,928	239 / 5,981	241 / 6,022	242 / 6,056	243 / 6,082	244 / 6,106	245 / 6,142	246 / 6,169	248 / 6,208	249 / 6,234	250 / 6,258	251 / 6,286	252 / 6,302	253 / 6,323	253 / 6,334	254 / 6,352	254 / 6,361	254 / 6,366
2	18.51 / 98.49	19.00 / 99.00	19.16 / 99.17	19.25 / 99.25	19.30 / 99.30	19.33 / 99.33	19.36 / 99.34	19.37 / 99.36	19.38 / 99.38	19.39 / 99.40	19.40 / 99.41	19.41 / 99.42	19.42 / 99.43	19.43 / 99.44	19.44 / 99.45	19.45 / 99.46	19.46 / 99.47	19.47 / 99.48	19.47 / 99.48	19.48 / 99.49	19.49 / 99.49	19.49 / 99.49	19.50 / 99.50	19.50 / 99.50
3	10.13 / 34.12	9.55 / 30.82	9.28 / 29.46	9.12 / 28.71	9.01 / 28.24	8.94 / 27.91	8.88 / 27.67	8.84 / 27.49	8.81 / 27.34	8.78 / 27.23	8.76 / 27.13	8.74 / 27.05	8.71 / 26.92	8.69 / 26.83	8.66 / 26.69	8.64 / 26.60	8.62 / 26.50	8.60 / 26.41	8.58 / 26.35	8.57 / 26.27	8.56 / 26.23	8.54 / 26.18	8.54 / 26.14	8.53 / 26.12
4	7.71 / 21.20	6.94 / 18.00	6.59 / 16.69	6.39 / 15.98	6.26 / 15.52	6.16 / 15.21	6.09 / 14.98	6.04 / 14.80	6.00 / 14.66	5.96 / 14.54	5.93 / 14.45	5.91 / 14.37	5.87 / 14.24	5.84 / 14.15	5.80 / 14.02	5.77 / 13.93	5.74 / 13.83	5.71 / 13.74	5.70 / 13.69	5.68 / 13.61	5.66 / 13.57	5.66 / 13.52	5.64 / 13.48	5.63 / 13.46
5	6.61 / 16.26	5.79 / 13.27	5.41 / 12.06	5.19 / 11.39	5.05 / 10.97	4.95 / 10.67	4.88 / 10.45	4.82 / 10.27	4.78 / 10.15	4.74 / 10.05	4.70 / 9.96	4.68 / 9.89	4.64 / 9.77	4.60 / 9.68	4.56 / 9.55	4.53 / 9.47	4.50 / 9.38	4.46 / 9.29	4.44 / 9.24	4.42 / 9.17	4.40 / 9.13	4.38 / 9.07	4.37 / 9.04	4.36 / 9.02
6	5.99 / 13.74	5.14 / 10.92	4.76 / 9.78	4.53 / 9.15	4.39 / 8.75	4.28 / 8.47	4.21 / 8.26	4.15 / 8.10	4.10 / 7.98	4.06 / 7.87	4.03 / 7.79	4.00 / 7.72	3.96 / 7.60	3.92 / 7.52	3.87 / 7.39	3.84 / 7.31	3.81 / 7.23	3.77 / 7.14	3.75 / 7.09	3.72 / 7.02	3.71 / 6.99	3.69 / 6.94	3.68 / 6.90	3.67 / 6.88
7	5.59 / 12.25	4.74 / 9.55	4.35 / 8.45	4.12 / 7.85	3.97 / 7.46	3.87 / 7.19	3.79 / 7.00	3.73 / 6.84	3.68 / 6.71	3.63 / 6.62	3.60 / 6.54	3.57 / 6.47	3.52 / 6.35	3.49 / 6.27	3.44 / 6.15	3.41 / 6.07	3.38 / 5.98	3.34 / 5.90	3.32 / 5.85	3.29 / 5.78	3.28 / 5.75	3.25 / 5.70	3.24 / 5.67	3.23 / 5.65
8	5.32 / 11.26	4.46 / 8.65	4.07 / 7.59	3.84 / 7.01	3.69 / 6.63	3.58 / 6.37	3.50 / 6.19	3.44 / 6.03	3.39 / 5.91	3.34 / 5.82	3.31 / 5.74	3.28 / 5.67	3.23 / 5.56	3.20 / 5.48	3.15 / 5.36	3.12 / 5.28	3.08 / 5.20	3.05 / 5.11	3.03 / 5.06	3.00 / 5.00	2.98 / 4.96	2.96 / 4.91	2.94 / 4.88	2.93 / 4.86
9	5.12 / 10.56	4.26 / 8.02	3.86 / 6.99	3.63 / 6.42	3.48 / 6.06	3.37 / 5.80	3.29 / 5.62	3.23 / 5.47	3.18 / 5.35	3.13 / 5.26	3.10 / 5.18	3.07 / 5.11	3.02 / 5.00	2.98 / 4.92	2.93 / 4.80	2.90 / 4.73	2.86 / 4.64	2.82 / 4.56	2.80 / 4.51	2.77 / 4.45	2.76 / 4.41	2.73 / 4.36	2.72 / 4.33	2.71 / 4.31
10	4.96 / 10.04	4.10 / 7.56	3.71 / 6.55	3.48 / 5.99	3.33 / 5.64	3.22 / 5.39	3.14 / 5.21	3.07 / 5.06	3.02 / 4.95	2.97 / 4.85	2.94 / 4.78	2.91 / 4.71	2.86 / 4.60	2.82 / 4.52	2.77 / 4.41	2.74 / 4.33	2.70 / 4.25	2.67 / 4.17	2.64 / 4.12	2.61 / 4.05	2.59 / 4.01	2.56 / 3.96	2.55 / 3.93	2.54 / 3.91
11	4.84 / 9.65	3.98 / 7.20	3.59 / 6.22	3.36 / 5.67	3.20 / 5.32	3.09 / 5.07	3.01 / 4.88	2.95 / 4.74	2.90 / 4.63	2.86 / 4.54	2.82 / 4.46	2.79 / 4.40	2.74 / 4.29	2.70 / 4.21	2.65 / 4.10	2.61 / 4.02	2.57 / 3.94	2.53 / 3.86	2.50 / 3.80	2.47 / 3.74	2.45 / 3.70	2.42 / 3.66	2.41 / 3.62	2.40 / 3.60
12	4.75 / 9.33	3.88 / 6.93	3.49 / 5.95	3.26 / 5.41	3.11 / 5.06	3.00 / 4.82	2.92 / 4.65	2.85 / 4.50	2.80 / 4.39	2.76 / 4.30	2.72 / 4.22	2.69 / 4.16	2.64 / 4.05	2.60 / 3.98	2.54 / 3.86	2.50 / 3.78	2.46 / 3.70	2.42 / 3.61	2.40 / 3.56	2.36 / 3.49	2.35 / 3.46	2.32 / 3.41	2.31 / 3.38	2.30 / 3.36
13	4.67 / 9.07	3.80 / 6.70	3.41 / 5.74	3.18 / 5.20	3.02 / 4.86	2.92 / 4.62	2.84 / 4.44	2.77 / 4.30	2.72 / 4.19	2.67 / 4.10	2.63 / 4.02	2.60 / 3.96	2.55 / 3.85	2.51 / 3.78	2.46 / 3.67	2.42 / 3.59	2.38 / 3.51	2.34 / 3.42	2.32 / 3.37	2.28 / 3.30	2.26 / 3.27	2.24 / 3.21	2.22 / 3.18	2.21 / 3.16

(continued)

* To be significant the F obtained from the data must be equal to or larger than the value shown in the table.
SOURCE: From Statistical Methods, by G. W. Snedecor and W. G. Cochran, Seventh Edition. Copyright © 1980 Iowa State University Press.

(continued)

Degrees of freedom (for the numerator of F ratio)

Degrees of freedom (for the denominator of F ratio)

	1	2	3	4	5	6	7	8	9	10	11	12	14	16	20	24	30	40	50	75	100	200	500	∞
14	4.60 / 8.86	3.74 / 6.51	3.34 / 5.56	3.11 / 5.03	2.96 / 4.69	2.85 / 4.46	2.77 / 4.28	2.70 / 4.14	2.65 / 4.03	2.60 / 3.94	2.56 / 3.86	2.53 / 3.80	2.48 / 3.70	2.44 / 3.62	2.39 / 3.51	2.35 / 3.43	2.31 / 3.34	2.27 / 3.26	2.24 / 3.21	2.21 / 3.14	2.19 / 3.11	2.16 / 3.06	2.14 / 3.02	2.13 / 3.00
15	4.54 / 8.68	3.68 / 6.36	3.29 / 5.52	3.06 / 4.89	2.90 / 4.56	2.79 / 4.32	2.70 / 4.14	2.64 / 4.00	2.59 / 3.89	2.55 / 3.80	2.51 / 3.73	2.48 / 3.67	2.43 / 3.56	2.39 / 3.48	2.33 / 3.36	2.29 / 3.29	2.25 / 3.20	2.21 / 3.12	2.18 / 3.07	2.15 / 3.00	2.12 / 2.97	2.10 / 2.92	2.08 / 2.89	2.07 / 2.87
16	4.49 / 8.53	3.63 / 6.23	3.24 / 5.29	3.01 / 4.77	2.85 / 4.44	2.74 / 4.20	2.66 / 4.03	2.59 / 3.89	2.54 / 3.78	2.49 / 3.69	2.45 / 3.61	2.42 / 3.55	2.37 / 3.45	2.33 / 3.37	2.28 / 3.25	2.24 / 3.18	2.20 / 3.10	2.16 / 3.01	2.13 / 2.96	2.09 / 2.89	2.07 / 2.86	2.04 / 2.80	2.02 / 2.77	2.01 / 2.75
17	4.45 / 8.40	3.59 / 6.11	3.20 / 5.18	2.96 / 4.67	2.81 / 4.34	2.70 / 4.10	2.62 / 3.93	2.55 / 3.79	2.50 / 3.68	2.45 / 3.59	2.41 / 3.52	2.38 / 3.45	2.33 / 3.35	2.29 / 3.27	2.23 / 3.16	2.19 / 3.08	2.15 / 3.00	2.11 / 2.92	2.08 / 2.86	2.04 / 2.79	2.02 / 2.76	1.99 / 2.70	1.97 / 2.67	1.96 / 2.65
18	4.41 / 8.28	3.55 / 6.01	3.16 / 5.09	2.93 / 4.58	2.77 / 4.25	2.66 / 4.01	2.58 / 3.85	2.51 / 3.71	2.46 / 3.60	2.41 / 3.51	2.37 / 3.44	2.34 / 3.37	2.29 / 3.27	2.25 / 3.19	2.19 / 3.07	2.15 / 3.00	2.11 / 2.91	2.07 / 2.83	2.04 / 2.78	2.00 / 2.71	1.98 / 2.68	1.95 / 2.62	1.93 / 2.59	1.92 / 2.57
19	4.38 / 8.18	3.52 / 5.93	3.13 / 5.01	2.90 / 4.50	2.74 / 4.17	2.63 / 3.94	2.55 / 3.77	2.48 / 3.63	2.43 / 3.52	2.38 / 3.43	2.34 / 3.36	2.31 / 3.30	2.26 / 3.19	2.21 / 3.12	2.15 / 3.00	2.11 / 2.92	2.07 / 2.84	2.02 / 2.76	2.00 / 2.70	1.96 / 2.63	1.94 / 2.60	1.91 / 2.54	1.90 / 2.51	1.88 / 2.49
20	4.35 / 8.10	3.49 / 5.85	3.10 / 4.94	2.87 / 4.43	2.71 / 4.10	2.60 / 3.87	2.52 / 3.71	2.45 / 3.56	2.40 / 3.45	2.35 / 3.37	2.31 / 3.30	2.28 / 3.23	2.23 / 3.13	2.18 / 3.05	2.12 / 2.94	2.08 / 2.86	2.04 / 2.77	1.99 / 2.69	1.96 / 2.63	1.92 / 2.56	1.90 / 2.53	1.87 / 2.47	1.85 / 2.44	1.84 / 2.42
21	4.32 / 8.02	3.47 / 5.78	3.07 / 4.87	2.84 / 4.37	2.68 / 4.04	2.57 / 3.81	2.49 / 3.65	2.42 / 3.51	2.37 / 3.40	2.32 / 3.31	2.28 / 3.24	2.25 / 3.17	2.20 / 3.07	2.15 / 2.99	2.09 / 2.88	2.05 / 2.80	2.00 / 2.72	1.96 / 2.63	1.93 / 2.58	1.89 / 2.51	1.87 / 2.47	1.84 / 2.42	1.82 / 2.38	1.81 / 2.36
22	4.30 / 7.94	3.44 / 5.72	3.05 / 4.82	2.82 / 4.31	2.66 / 3.99	2.55 / 3.76	2.47 / 3.59	2.40 / 3.45	2.35 / 3.35	2.30 / 3.26	2.26 / 3.18	2.23 / 3.12	2.18 / 3.02	2.13 / 2.94	2.07 / 2.83	2.03 / 2.75	1.98 / 2.67	1.93 / 2.58	1.91 / 2.53	1.87 / 2.46	1.84 / 2.42	1.81 / 2.37	1.80 / 2.33	1.78 / 2.31
23	4.28 / 7.88	3.42 / 5.66	3.03 / 4.76	2.80 / 4.26	2.64 / 3.94	2.53 / 3.71	2.45 / 3.54	2.38 / 3.41	2.32 / 3.30	2.28 / 3.21	2.24 / 3.14	2.20 / 3.07	2.14 / 2.97	2.10 / 2.89	2.04 / 2.78	2.00 / 2.70	1.96 / 2.62	1.91 / 2.53	1.88 / 2.48	1.84 / 2.41	1.82 / 2.37	1.79 / 2.32	1.77 / 2.28	1.76 / 2.26
24	4.26 / 7.82	3.40 / 5.61	3.01 / 4.72	2.78 / 4.22	2.62 / 3.90	2.51 / 3.67	2.43 / 3.50	2.36 / 3.36	2.30 / 3.25	2.26 / 3.17	2.22 / 3.09	2.18 / 3.03	2.13 / 2.93	2.09 / 2.85	2.02 / 2.74	1.98 / 2.66	1.94 / 2.58	1.89 / 2.49	1.86 / 2.44	1.82 / 2.36	1.80 / 2.33	1.76 / 2.27	1.74 / 2.23	1.73 / 2.21
25	4.24 / 7.77	3.38 / 5.57	2.99 / 4.68	2.76 / 4.18	2.60 / 3.86	2.49 / 3.63	2.41 / 3.46	2.34 / 3.32	2.28 / 3.21	2.24 / 3.13	2.20 / 3.05	2.16 / 2.99	2.11 / 2.89	2.06 / 2.81	2.00 / 2.70	1.96 / 2.62	1.92 / 2.54	1.87 / 2.45	1.84 / 2.40	1.80 / 2.32	1.77 / 2.29	1.74 / 2.23	1.72 / 2.19	1.71 / 2.17
26	4.22 / 7.72	3.37 / 5.53	2.98 / 4.64	2.74 / 4.14	2.59 / 3.82	2.47 / 3.59	2.39 / 3.42	2.32 / 3.29	2.27 / 3.17	2.22 / 3.09	2.18 / 3.02	2.15 / 2.96	2.10 / 2.86	2.05 / 2.77	1.99 / 2.66	1.95 / 2.58	1.90 / 2.50	1.85 / 2.41	1.82 / 2.36	1.78 / 2.28	1.76 / 2.25	1.72 / 2.19	1.70 / 2.15	1.69 / 2.13

(continued)

(continued)

(continued)

Degrees of freedom (for the numerator of F ratio)

Degrees of freedom (for the denominator of F ratio)

	1	2	3	4	5	6	7	8	9	10	11	12	14	16	20	24	30	40	50	75	100	200	500	∞	
27	4.21	3.35	2.96	2.73	2.57	2.46	2.37	2.30	2.25	2.20	2.16	2.13	2.08	2.03	1.97	1.93	1.88	1.84	1.80	1.76	1.74	1.71	1.68	1.67	27
	7.68	5.49	4.60	4.11	3.79	3.56	3.39	3.26	3.14	3.06	2.98	2.93	2.83	2.74	2.63	2.55	2.47	2.38	2.33	2.25	2.21	2.16	2.12	2.10	
28	4.20	3.34	2.95	2.71	2.56	2.44	2.36	2.29	2.24	2.19	2.15	2.12	2.06	2.02	1.96	1.91	1.87	1.81	1.78	1.75	1.72	1.69	1.67	1.65	28
	7.64	5.45	4.57	4.07	3.76	3.53	3.36	3.23	3.11	3.03	2.95	2.90	2.80	2.71	2.60	2.52	2.44	2.35	2.30	2.22	2.18	2.13	2.09	2.06	
29	4.18	3.33	2.93	2.70	2.54	2.43	2.35	2.28	2.22	2.18	2.14	2.10	2.05	2.00	1.94	1.90	1.85	1.80	1.77	1.73	1.71	1.68	1.65	1.64	29
	7.60	5.42	4.54	4.04	3.73	3.50	3.33	3.20	3.08	3.00	2.92	2.87	2.77	2.68	2.57	2.49	2.41	2.32	2.27	2.19	2.15	2.10	2.06	2.03	
30	4.17	3.32	2.92	2.69	2.53	2.42	2.34	2.27	2.21	2.16	2.12	2.09	2.04	1.99	1.93	1.89	1.84	1.79	1.76	1.72	1.69	1.66	1.64	1.62	30
	7.56	5.39	4.51	4.02	3.70	3.47	3.30	3.17	3.06	2.98	2.90	2.84	2.74	2.66	2.55	2.47	2.38	2.29	2.24	2.16	2.13	2.07	2.03	2.01	
32	4.15	3.30	2.90	2.67	2.51	2.40	2.32	2.25	2.19	2.14	2.10	2.07	2.02	1.97	1.91	1.86	1.82	1.76	1.74	1.69	1.67	1.64	1.61	1.59	32
	7.50	5.34	4.46	3.97	3.66	3.42	3.25	3.12	3.01	2.94	2.86	2.80	2.70	2.62	2.51	2.42	2.34	2.25	2.20	2.12	2.08	2.02	1.98	1.96	
34	4.13	3.28	2.88	2.65	2.49	2.38	2.30	2.23	2.17	2.12	2.08	2.05	2.00	1.95	1.89	1.84	1.80	1.74	1.71	1.67	1.64	1.61	1.59	1.57	34
	7.44	5.29	4.42	3.93	3.61	3.38	3.21	3.08	2.97	2.89	2.82	2.76	2.66	2.58	2.47	2.38	2.30	2.21	2.15	2.08	2.04	1.98	1.94	1.91	
36	4.11	3.26	2.86	2.63	2.48	2.36	2.28	2.21	2.15	2.10	2.06	2.03	1.98	1.93	1.87	1.82	1.78	1.72	1.69	1.65	1.62	1.59	1.56	1.55	36
	7.39	5.25	4.38	3.89	3.58	3.35	3.18	3.04	2.94	2.86	2.78	2.72	2.62	2.54	2.43	2.35	2.26	2.17	2.12	2.04	2.00	1.94	1.90	1.87	
38	4.10	3.25	2.85	2.62	2.46	2.35	2.26	2.19	2.14	2.09	2.05	2.02	1.96	1.92	1.85	1.80	1.76	1.71	1.67	1.63	1.60	1.57	1.54	1.53	38
	7.35	5.21	4.34	3.86	3.54	3.32	3.15	3.02	2.91	2.82	2.75	2.69	2.59	2.51	2.40	2.32	2.22	2.14	2.08	2.00	1.97	1.90	1.86	1.84	
40	4.08	3.23	2.84	2.61	2.45	2.34	2.25	2.18	2.12	2.07	2.04	2.00	1.95	1.90	1.84	1.79	1.74	1.69	1.66	1.61	1.59	1.55	1.53	1.51	40
	7.31	5.18	4.31	3.83	3.51	3.29	3.12	2.99	2.88	2.80	2.73	2.66	2.56	2.49	2.37	2.29	2.20	2.11	2.05	1.97	1.94	1.88	1.84	1.81	
42	4.07	3.22	2.83	2.59	2.44	2.32	2.24	2.17	2.11	2.06	2.02	1.99	1.94	1.89	1.82	1.78	1.73	1.68	1.64	1.60	1.57	1.54	1.51	1.49	42
	7.27	5.15	4.29	3.80	3.49	3.26	3.10	2.96	2.86	2.77	2.70	2.64	2.54	2.46	2.35	2.26	2.17	2.08	2.02	1.94	1.91	1.85	1.80	1.78	
44	4.06	3.21	2.82	2.58	2.43	2.31	2.23	2.16	2.10	2.05	2.01	1.98	1.92	1.88	1.81	1.76	1.72	1.66	1.63	1.58	1.56	1.52	1.50	1.48	44
	7.24	5.12	4.26	3.78	3.46	3.24	3.07	2.94	2.84	2.75	2.68	2.62	2.52	2.44	2.32	2.24	2.15	2.06	2.00	1.92	1.88	1.82	1.78	1.75	
46	4.05	3.20	2.81	2.57	2.42	2.30	2.22	2.14	2.09	2.04	2.00	1.97	1.91	1.87	1.80	1.75	1.71	1.65	1.62	1.57	1.54	1.51	1.48	1.46	46
	7.21	5.10	4.24	3.76	3.44	3.22	3.05	2.92	2.82	2.73	2.66	2.60	2.50	2.42	2.30	2.22	2.13	2.04	1.98	1.90	1.86	1.80	1.76	1.72	
48	4.04	3.19	2.80	2.56	2.41	2.30	2.21	2.14	2.08	2.03	1.99	1.96	1.90	1.86	1.79	1.74	1.70	1.64	1.61	1.56	1.53	1.50	1.47	1.45	48
	7.19	5.08	4.22	3.74	3.42	3.20	3.04	2.90	2.80	2.71	2.64	2.58	2.48	2.40	2.28	2.20	2.11	2.02	1.96	1.88	1.84	1.78	1.73	1.70	

(continued)

(continued)

	Degrees of freedom (for the numerator of F ratio)

Degrees of freedom (for the denominator of F ratio)	1	2	3	4	5	6	7	8	9	10	11	12	14	16	20	24	30	40	50	75	100	200	500	∞
50	4.03 / 7.17	3.18 / 5.06	2.79 / 4.20	2.56 / 3.72	2.40 / 3.41	2.29 / 3.18	2.20 / 3.02	2.13 / 2.88	2.07 / 2.78	2.02 / 2.70	1.98 / 2.62	1.95 / 2.56	1.90 / 2.46	1.85 / 2.39	1.78 / 2.26	1.74 / 2.18	1.69 / 2.10	1.63 / 2.00	1.60 / 1.94	1.55 / 1.86	1.52 / 1.82	1.48 / 1.76	1.46 / 1.71	1.44 / 1.68
55	4.02 / 7.12	3.17 / 5.01	2.78 / 4.16	2.54 / 3.68	2.38 / 3.37	2.27 / 3.15	2.18 / 2.98	2.11 / 2.85	2.05 / 2.75	2.00 / 2.66	1.97 / 2.59	1.93 / 2.53	1.88 / 2.43	1.83 / 2.35	1.76 / 2.23	1.72 / 2.15	1.67 / 2.06	1.61 / 1.96	1.58 / 1.90	1.52 / 1.82	1.50 / 1.78	1.46 / 1.71	1.43 / 1.66	1.41 / 1.64
60	4.00 / 7.08	3.15 / 4.98	2.76 / 4.13	2.52 / 3.65	2.37 / 3.34	2.25 / 3.12	2.17 / 2.95	2.10 / 2.82	2.04 / 2.72	1.99 / 2.63	1.95 / 2.56	1.92 / 2.50	1.86 / 2.40	1.81 / 2.32	1.75 / 2.20	1.70 / 2.12	1.65 / 2.03	1.59 / 1.93	1.56 / 1.87	1.50 / 1.79	1.48 / 1.74	1.44 / 1.68	1.41 / 1.63	1.39 / 1.60
65	3.99 / 7.04	3.14 / 4.95	2.75 / 4.10	2.51 / 3.62	2.36 / 3.31	2.24 / 3.09	2.15 / 2.93	2.08 / 2.79	2.02 / 2.70	1.98 / 2.61	1.94 / 2.54	1.90 / 2.47	1.85 / 2.37	1.80 / 2.30	1.73 / 2.18	1.68 / 2.09	1.63 / 2.00	1.57 / 1.90	1.54 / 1.84	1.49 / 1.76	1.46 / 1.71	1.42 / 1.64	1.39 / 1.60	1.37 / 1.56
70	3.98 / 7.01	3.13 / 4.92	2.74 / 4.08	2.50 / 3.60	2.35 / 3.29	2.23 / 3.07	2.14 / 2.91	2.07 / 2.77	2.01 / 2.67	1.97 / 2.59	1.93 / 2.51	1.89 / 2.45	1.84 / 2.35	1.79 / 2.28	1.72 / 2.15	1.67 / 2.07	1.62 / 1.98	1.56 / 1.88	1.53 / 1.82	1.47 / 1.74	1.45 / 1.69	1.40 / 1.62	1.37 / 1.56	1.35 / 1.53
80	3.96 / 6.96	3.11 / 4.88	2.72 / 4.04	2.48 / 3.56	2.33 / 3.25	2.21 / 3.04	2.12 / 2.87	2.05 / 2.74	1.99 / 2.64	1.95 / 2.55	1.91 / 2.48	1.88 / 2.41	1.82 / 2.32	1.77 / 2.24	1.70 / 2.11	1.65 / 2.03	1.60 / 1.94	1.54 / 1.84	1.51 / 1.78	1.45 / 1.70	1.42 / 1.65	1.38 / 1.57	1.35 / 1.52	1.32 / 1.49
100	3.94 / 6.90	3.09 / 4.82	2.70 / 3.98	2.46 / 3.51	2.30 / 3.20	2.19 / 2.99	2.10 / 2.82	2.03 / 2.69	1.97 / 2.59	1.92 / 2.51	1.88 / 2.43	1.85 / 2.36	1.79 / 2.26	1.75 / 2.19	1.68 / 2.06	1.63 / 1.98	1.57 / 1.89	1.51 / 1.79	1.48 / 1.73	1.42 / 1.64	1.39 / 1.59	1.34 / 1.51	1.30 / 1.46	1.28 / 1.43
125	3.92 / 6.84	3.07 / 4.78	2.68 / 3.94	2.44 / 3.47	2.29 / 3.17	2.17 / 2.95	2.08 / 2.79	2.01 / 2.65	1.95 / 2.56	1.90 / 2.47	1.86 / 2.40	1.83 / 2.33	1.77 / 2.23	1.72 / 2.15	1.65 / 2.03	1.60 / 1.94	1.55 / 1.85	1.49 / 1.75	1.45 / 1.68	1.39 / 1.59	1.36 / 1.54	1.31 / 1.46	1.27 / 1.40	1.25 / 1.37
150	3.91 / 6.81	3.06 / 4.75	2.67 / 3.91	2.43 / 3.44	2.27 / 3.14	2.16 / 2.92	2.07 / 2.76	2.00 / 2.62	1.94 / 2.53	1.89 / 2.44	1.85 / 2.37	1.82 / 2.30	1.76 / 2.20	1.71 / 2.12	1.64 / 2.00	1.59 / 1.91	1.54 / 1.83	1.47 / 1.72	1.44 / 1.66	1.37 / 1.56	1.34 / 1.51	1.29 / 1.43	1.25 / 1.37	1.22 / 1.33
200	3.89 / 6.76	3.04 / 4.71	2.65 / 3.88	2.41 / 3.41	2.26 / 3.11	2.14 / 2.90	2.05 / 2.73	1.98 / 2.60	1.92 / 2.50	1.87 / 2.41	1.83 / 2.34	1.80 / 2.28	1.74 / 2.17	1.69 / 2.09	1.62 / 1.97	1.57 / 1.88	1.52 / 1.79	1.45 / 1.69	1.42 / 1.62	1.35 / 1.53	1.32 / 1.48	1.26 / 1.39	1.22 / 1.33	1.19 / 1.28
400	3.86 / 6.70	3.02 / 4.66	2.62 / 3.83	2.39 / 3.36	2.23 / 3.06	2.12 / 2.85	2.03 / 2.69	1.96 / 2.55	1.90 / 2.46	1.85 / 2.37	1.81 / 2.29	1.78 / 2.23	1.72 / 2.12	1.67 / 2.04	1.60 / 1.92	1.54 / 1.84	1.49 / 1.74	1.42 / 1.64	1.38 / 1.57	1.32 / 1.47	1.28 / 1.42	1.22 / 1.32	1.16 / 1.24	1.13 / 1.19
1000	3.85 / 6.66	3.00 / 4.62	2.61 / 3.80	2.38 / 3.34	2.22 / 3.04	2.10 / 2.82	2.02 / 2.66	1.95 / 2.53	1.89 / 2.43	1.84 / 2.34	1.80 / 2.26	1.76 / 2.20	1.70 / 2.09	1.65 / 2.01	1.58 / 1.89	1.53 / 1.81	1.47 / 1.71	1.41 / 1.61	1.36 / 1.54	1.30 / 1.44	1.26 / 1.38	1.19 / 1.28	1.13 / 1.19	1.08 / 1.11
∞	3.84 / 6.64	2.99 / 4.60	2.60 / 3.78	2.37 / 3.32	2.21 / 3.02	2.09 / 2.80	2.01 / 2.64	1.94 / 2.51	1.88 / 2.41	1.83 / 2.32	1.79 / 2.24	1.75 / 2.18	1.69 / 2.07	1.64 / 1.99	1.57 / 1.87	1.52 / 1.79	1.46 / 1.69	1.40 / 1.59	1.35 / 1.52	1.28 / 1.41	1.24 / 1.36	1.17 / 1.25	1.11 / 1.15	1.00 / 1.00

Tukey HSD test table

	Number of levels of the independent variable													
df_{wg}	2	3	4	5	6	7	8	9	10	11	12	13	14	15
1	17.97	26.98	32.82	37.08	40.41	43.12	45.40	47.36	49.07	50.59	51.96	53.20	54.33	55.36
2	6.08	8.33	9.80	10.88	11.74	12.44	13.03	13.54	13.99	14.39	14.75	15.08	15.38	15.65
3	4.50	5.91	6.82	7.50	8.04	8.48	8.85	9.18	9.46	9.72	9.95	10.15	10.35	10.53
4	3.93	5.04	5.76	6.29	6.71	7.05	7.35	7.60	7.83	8.03	8.21	8.37	8.52	8.66
5	3.64	4.60	5.22	5.67	6.03	6.33	6.58	6.80	7.00	7.17	7.32	7.47	7.60	7.72
6	3.46	4.34	4.90	5.31	5.63	5.90	6.12	6.32	6.49	6.65	6.79	6.92	7.03	7.14
7	3.34	4.16	4.68	5.06	5.36	5.61	5.82	6.00	6.16	6.30	6.43	6.55	6.66	6.76
8	3.26	4.04	4.53	4.89	5.17	5.40	5.60	5.77	5.92	6.05	6.18	6.29	6.39	6.48
9	3.20	3.95	4.42	4.76	5.02	5.24	5.43	5.60	5.74	5.87	5.98	6.09	6.19	6.28
10	3.15	3.88	4.33	4.65	4.91	5.12	5.30	5.46	5.60	5.72	5.83	5.94	6.03	6.11
11	3.11	3.82	4.26	4.57	4.82	5.03	5.20	5.35	5.49	5.60	5.71	5.81	5.90	5.98
12	3.08	3.77	4.20	4.51	4.75	4.95	5.12	5.26	5.40	5.51	5.62	5.71	5.79	5.88
13	3.06	3.74	4.15	4.45	4.69	4.88	5.05	5.19	5.32	5.43	5.53	5.63	5.71	5.79
14	3.03	3.70	4.11	4.41	4.64	4.83	4.99	5.13	5.25	5.36	5.46	5.55	5.64	5.71
15	3.01	3.67	4.08	4.37	4.60	4.78	4.94	5.08	5.20	5.31	5.40	5.49	5.57	5.65
16	3.00	3.65	4.05	4.33	4.56	4.74	4.90	5.03	5.15	5.26	5.35	5.44	5.52	5.59
17	2.98	3.63	4.02	4.30	4.52	4.70	4.86	4.99	5.11	5.21	5.31	5.39	5.47	5.54
18	2.97	3.61	4.00	4.28	4.50	4.67	4.82	4.96	5.07	5.17	5.27	5.35	5.43	5.50
19	2.96	3.59	3.98	4.25	4.47	4.64	4.79	4.92	5.04	5.14	5.23	5.32	5.39	5.46
20	2.95	3.58	3.96	4.23	4.44	4.62	4.77	4.90	5.01	5.11	5.20	5.28	5.36	5.43
24	2.92	3.53	3.90	4.17	4.37	4.54	4.68	4.81	4.92	5.01	5.10	5.18	5.25	5.32
30	2.89	3.49	3.84	4.10	4.30	4.46	4.60	4.72	4.82	4.92	5.00	5.08	5.15	5.21
40	2.86	3.44	3.79	4.04	4.23	4.39	4.52	4.64	4.74	4.82	4.90	4.98	5.04	5.11
60	2.83	3.40	3.74	3.98	4.16	4.31	4.44	4.55	4.65	4.73	4.81	4.88	4.94	5.00
120	2.80	3.36	3.69	3.92	4.10	4.24	4.36	4.47	4.56	4.64	4.71	4.78	4.84	4.90
∞	2.77	3.31	3.63	3.86	4.03	4.17	4.29	4.39	4.47	4.55	4.62	4.68	4.74	4.80

SOURCE: From "Tables of Range and Studentized Range," by M. L. Harter, 1960, *Annals of Mathematical Statistics, 31,* 1122–1147. Copyright © 1960 The Institute of Mathematical Statistics. Reprinted with permission.

(*continued*)

(continued)

α = .01

df_{ws}	2	3	4	5	6	7	8	9	10	11	12	13	14	15
1	90.03	135.00	164.30	185.60	202.20	215.80	227.20	237.00	245.60	253.20	260.00	266.20	271.80	277.00
2	14.04	19.02	22.29	24.72	26.63	28.20	29.53	30.68	31.69	32.59	33.40	34.13	34.81	35.43
3	8.26	10.62	12.17	13.33	14.24	15.00	15.64	16.20	16.69	17.13	17.53	17.89	18.22	18.52
4	6.51	8.12	9.17	9.96	10.58	11.10	11.55	11.93	12.27	12.57	12.84	13.09	13.32	13.53
5	5.70	6.98	7.80	8.42	8.91	9.32	9.67	9.97	10.24	10.48	10.70	10.89	11.08	11.24
6	5.24	6.33	7.03	7.56	7.97	8.32	8.62	8.87	9.10	9.30	9.48	9.65	9.81	9.95
7	4.95	5.92	6.54	7.00	7.37	7.68	7.94	8.17	8.37	8.55	8.71	8.86	9.00	9.12
8	4.75	5.64	6.20	6.62	6.96	7.24	7.47	7.68	7.86	8.03	8.18	8.31	8.44	8.55
9	4.60	5.43	5.96	6.35	6.66	6.92	7.13	7.32	7.50	7.65	7.78	7.91	8.02	8.13
10	4.48	5.27	5.77	6.14	6.43	6.67	6.88	7.06	7.21	7.36	7.48	7.60	7.71	7.81
11	4.39	5.15	5.62	5.97	6.25	6.48	6.67	6.84	6.99	7.13	7.25	7.36	7.46	7.56
12	4.32	5.05	5.50	5.84	6.10	6.32	6.51	6.67	6.81	6.94	7.06	7.17	7.26	7.36
13	4.26	4.96	5.40	5.73	5.98	6.19	6.37	6.53	6.67	6.79	6.90	7.01	7.10	7.19
14	4.21	4.90	5.32	5.63	5.88	6.08	6.26	6.41	6.54	6.66	6.77	6.87	6.96	7.05
15	4.17	4.84	5.25	5.56	5.80	5.99	6.16	6.31	6.44	6.56	6.66	6.76	6.84	6.93
16	4.13	4.79	5.19	5.49	5.72	5.92	6.08	6.22	6.35	6.46	6.56	6.66	6.74	6.82
17	4.10	4.74	5.14	5.43	5.66	5.85	6.01	6.15	6.27	6.38	6.48	6.57	6.66	6.73
18	4.07	4.70	5.09	5.38	5.60	5.79	5.94	6.08	6.20	6.31	6.41	6.50	6.58	6.66
19	4.05	4.67	5.05	5.33	5.55	5.74	5.89	6.02	6.14	6.25	6.34	6.43	6.51	6.58
20	4.02	4.64	5.02	5.29	5.51	5.69	5.84	5.97	6.09	6.19	6.28	6.37	6.45	6.52
24	3.96	4.55	4.91	5.17	5.37	5.54	5.69	5.81	5.92	6.02	6.11	6.19	6.26	6.33
30	3.89	4.46	4.80	5.05	5.24	5.40	5.54	5.65	5.76	5.85	5.93	6.01	6.08	6.14
40	3.82	4.37	4.70	4.93	5.11	5.26	5.39	5.50	5.60	5.69	5.76	5.84	5.90	5.96
60	3.76	4.28	4.60	4.82	4.99	5.13	5.25	5.36	5.45	5.53	5.60	5.67	5.73	5.78
120	3.70	4.20	4.50	4.71	4.87	5.01	5.12	5.21	5.30	5.38	5.44	5.51	5.56	5.61
∞	3.64	4.12	4.40	4.60	4.76	4.88	4.99	5.08	5.16	5.23	5.29	5.35	5.40	5.45

Chi square Table

df	.10	.05	.02	.01	.001
			α levels		
1	2.71	3.84	5.41	6.64	10.83
2	4.60	5.99	7.82	9.21	13.82
3	6.25	7.82	9.84	11.34	16.27
4	7.78	9.49	11.67	13.28	18.46
5	9.24	11.07	13.39	15.09	20.52
6	10.64	12.59	15.03	16.81	22.46
7	12.02	14.07	16.62	18.48	24.32
8	13.36	15.51	18.17	20.09	26.12
9	14.68	16.92	19.68	21.67	27.88
10	15.99	18.31	21.16	23.21	29.59
11	17.28	19.68	22.62	24.72	31.26
12	18.55	21.03	24.05	26.22	32.91
13	19.81	22.36	25.47	27.69	34.53
14	21.06	23.68	26.87	29.14	36.12
15	22.31	25.00	28.26	30.58	37.70
16	23.54	26.30	29.63	32.00	39.25
17	24.77	27.59	31.00	33.41	40.79
18	25.99	28.87	32.35	34.80	42.31
19	27.20	30.14	33.69	36.19	43.82
20	28.41	31.41	35.02	37.57	45.32
21	29.62	32.67	36.34	38.93	46.80
22	30.81	33.92	37.66	40.29	48.27
23	32.01	35.17	38.97	41.64	49.73
24	33.20	36.42	40.27	42.98	51.18
25	34.38	37.65	41.57	44.31	52.62
26	35.56	38.88	42.86	45.64	54.05
27	36.74	40.11	44.14	46.96	55.48
28	37.92	41.34	45.42	48.28	56.89
29	39.09	42.56	46.69	49.59	58.30
30	40.26	43.77	47.96	50.89	59.70

* To be significant the χ^2 obtained from the data must be equal to or larger than the value shown in the table.

SOURCE: From Table IV in R. A. Fisher and F. Yates, *Statistical Tables for Biological, Agricultural, and Medical Research*, Sixth Edition, published by Addison Wesley Longman Ltd., (1974).

Mann-Whitney U test table

One-tailed test
$\alpha = .01$ (lightface)
$\alpha = .005$ (boldface)

Two-tailed test
$\alpha = .02$ (lightface)
$\alpha = .01$ (boldface)

N_2 \ N_1	1	2	3	4	5	6	7	8	9	10	11	12	13	14	15	16	17	18	19	20
1	—	—	—	—	—	—	—	—	—	—	—	—	—	—	—	—	—	—	—	—
2	—	—	—	—	—	—	—	—	—	—	—	—	0	0	0	0	0	0	1	1
	—	—	—	—	—	—	—	—	—	—	—	—	—	—	—	—	—	—	**0**	**0**
3	—	—	—	—	—	—	0	0	1	1	1	2	2	2	3	3	4	4	4	5
	—	—	—	—	—	—	—	—	**0**	**0**	**0**	**1**	**1**	**1**	**2**	**2**	**2**	**2**	**3**	**3**
4	—	—	—	—	0	1	1	2	3	3	4	5	5	6	7	7	8	9	9	10
	—	—	—	—	—	**0**	**0**	**1**	**1**	**2**	**2**	**3**	**3**	**4**	**5**	**5**	**6**	**6**	**7**	**8**
5	—	—	—	0	1	2	3	4	5	6	7	8	9	10	11	12	13	14	15	16
	—	—	—	—	**0**	**1**	**1**	**2**	**3**	**4**	**5**	**6**	**7**	**7**	**8**	**9**	**10**	**11**	**12**	**13**
6	—	—	—	1	2	3	4	6	7	8	9	11	12	13	15	16	18	19	20	22
	—	—	—	**0**	**1**	**2**	**3**	**4**	**5**	**6**	**7**	**9**	**10**	**11**	**12**	**13**	**15**	**16**	**17**	**18**
7	—	—	0	1	3	4	6	7	9	11	12	14	16	17	19	21	23	24	26	28
	—	—	—	**0**	**1**	**3**	**4**	**6**	**7**	**9**	**10**	**12**	**13**	**15**	**16**	**18**	**19**	**21**	**22**	**24**
8	—	—	0	2	4	6	7	9	11	13	15	17	20	22	24	26	28	30	32	34
	—	—	—	**1**	**2**	**4**	**6**	**7**	**9**	**11**	**13**	**15**	**17**	**18**	**20**	**22**	**24**	**26**	**29**	**30**
9	—	—	1	3	5	7	9	11	14	16	18	21	23	26	28	31	33	36	38	40
	—	—	**0**	**1**	**3**	**5**	**7**	**9**	**11**	**13**	**16**	**18**	**20**	**22**	**24**	**27**	**29**	**31**	**33**	**36**
10	—	—	1	3	6	8	11	13	16	19	22	24	27	30	33	36	38	41	44	47
	—	—	**0**	**2**	**4**	**6**	**9**	**11**	**13**	**16**	**18**	**21**	**24**	**26**	**29**	**31**	**34**	**37**	**39**	**42**
11	—	—	1	4	7	9	12	15	18	22	25	28	31	34	37	41	44	47	50	53
	—	—	**0**	**2**	**5**	**7**	**10**	**13**	**16**	**18**	**21**	**24**	**27**	**30**	**33**	**36**	**39**	**42**	**45**	**48**
12	—	—	2	5	8	11	14	17	21	24	28	31	35	38	42	46	49	53	56	60
	—	—	**1**	**3**	**6**	**9**	**12**	**15**	**18**	**21**	**24**	**27**	**31**	**34**	**37**	**41**	**44**	**47**	**51**	**54**
13	—	0	2	5	9	12	16	20	23	27	31	35	39	43	47	51	55	59	63	67
	—	—	**1**	**3**	**7**	**10**	**13**	**17**	**20**	**24**	**27**	**31**	**34**	**38**	**42**	**45**	**49**	**53**	**56**	**60**
14	—	0	2	6	10	13	17	22	26	30	34	38	43	47	51	56	60	65	69	73
	—	—	**1**	**4**	**7**	**11**	**15**	**18**	**22**	**26**	**30**	**34**	**38**	**42**	**46**	**50**	**54**	**58**	**63**	**67**
15	—	0	3	7	11	15	19	24	28	33	37	42	47	51	56	61	66	70	75	80
	—	—	**2**	**5**	**8**	**12**	**16**	**20**	**24**	**29**	**33**	**37**	**42**	**46**	**51**	**55**	**60**	**64**	**69**	**73**
16	—	0	3	7	12	16	21	26	31	36	41	46	51	56	61	66	71	76	82	87
	—	—	**2**	**5**	**9**	**13**	**18**	**22**	**27**	**31**	**36**	**41**	**45**	**50**	**55**	**60**	**65**	**70**	**74**	**79**
17	—	0	4	8	13	18	23	28	33	38	44	49	55	60	66	71	77	82	88	93
	—	—	**2**	**6**	**10**	**15**	**19**	**24**	**29**	**34**	**39**	**44**	**49**	**54**	**60**	**65**	**70**	**75**	**81**	**86**
18	—	0	4	9	14	19	24	30	36	41	47	53	59	65	70	76	82	88	94	100
	—	—	**2**	**6**	**11**	**16**	**21**	**26**	**31**	**37**	**42**	**47**	**53**	**58**	**64**	**70**	**75**	**81**	**87**	**92**
19	—	1	4	9	15	20	26	32	38	44	50	56	63	69	75	82	88	94	101	107
	—	**0**	**3**	**7**	**12**	**17**	**22**	**28**	**33**	**39**	**45**	**51**	**56**	**63**	**69**	**74**	**81**	**87**	**93**	**99**
20	—	1	5	10	16	22	28	34	40	47	53	60	67	73	80	87	93	100	107	114
	—	**0**	**3**	**8**	**13**	**18**	**24**	**30**	**36**	**42**	**48**	**54**	**60**	**67**	**73**	**79**	**86**	**92**	**99**	**105**

(continued)

* To be significant the U obtained from data must be equal to or **less than** the value shown in the table. Dashes in the body of the table indicate that no decision is possible at the stated level of significance.

SOURCE: From *Elementary Statistics*, Second Edition, by R. E. Kirk, Brooks/Cole Publishing, 1984.

(continued)

One-tailed test	Two-tailed test
$\alpha = .05$ (lightface)	$\alpha = .10$ (lightface)
$\alpha = .025$ (boldface)	$\alpha = .05$ (boldface)

N_2 \ N_1	1	2	3	4	5	6	7	8	9	10	11	12	13	14	15	16	17	18	19	20
1	—	—	—	—	—	—	—	—	—	—	—	—	—	—	—	—	—	—	0	0
2	—	—	—	—	0	0	0	1	1	1	1	2	2	2	3	3	3	4	4	4
	—	—	—	—	—	—	—	**0**	**0**	**0**	**0**	**1**	**1**	**1**	**1**	**1**	**2**	**2**	**2**	**2**
3	—	—	0	0	1	2	2	3	3	4	5	5	6	7	7	8	9	9	10	11
	—	—	—	—	**0**	**1**	**1**	**2**	**2**	**3**	**3**	**4**	**4**	**5**	**5**	**6**	**6**	**7**	**7**	**8**
4	—	—	0	1	2	3	4	5	6	7	8	9	10	11	12	14	15	16	17	18
	—	—	—	**0**	**1**	**2**	**3**	**4**	**4**	**5**	**6**	**7**	**8**	**9**	**10**	**11**	**11**	**12**	**13**	**13**
5	—	0	1	2	4	5	6	8	9	11	12	13	15	16	18	19	20	22	23	25
	—	—	**0**	**1**	**2**	**3**	**5**	**6**	**7**	**8**	**9**	**11**	**12**	**13**	**14**	**15**	**17**	**18**	**19**	**20**
6	—	0	2	3	5	7	8	10	12	14	16	17	19	21	23	25	26	28	30	32
	—	—	**1**	**2**	**3**	**5**	**6**	**8**	**10**	**11**	**13**	**14**	**16**	**17**	**19**	**21**	**22**	**24**	**25**	**27**
7	—	0	2	4	6	8	11	13	15	17	19	21	24	26	28	30	33	35	37	39
	—	—	**1**	**3**	**5**	**6**	**8**	**10**	**12**	**14**	**16**	**18**	**20**	**22**	**24**	**26**	**28**	**30**	**32**	**34**
8	—	1	3	5	8	10	13	15	18	20	23	26	28	31	33	36	39	41	44	47
	—	**0**	**2**	**4**	**6**	**8**	**10**	**13**	**15**	**17**	**19**	**22**	**24**	**26**	**29**	**31**	**34**	**36**	**38**	**41**
9	—	1	3	6	9	12	15	18	21	24	27	30	33	36	39	42	45	48	51	54
	—	**0**	**2**	**4**	**7**	**10**	**12**	**15**	**17**	**20**	**23**	**26**	**28**	**31**	**34**	**37**	**39**	**42**	**45**	**48**
10	—	1	4	7	11	14	17	20	24	27	31	34	37	41	44	48	51	55	58	62
	—	**0**	**3**	**5**	**8**	**11**	**14**	**17**	**20**	**23**	**26**	**29**	**33**	**36**	**39**	**42**	**45**	**48**	**52**	**55**
11	—	1	5	8	12	16	19	23	27	31	34	38	42	46	50	54	57	61	65	69
	—	**0**	**3**	**6**	**9**	**13**	**16**	**19**	**23**	**26**	**30**	**33**	**37**	**40**	**44**	**47**	**51**	**55**	**58**	**62**
12	—	2	5	9	13	17	21	26	30	34	38	42	47	51	55	60	64	68	72	77
	—	**1**	**4**	**7**	**11**	**14**	**18**	**22**	**26**	**29**	**33**	**37**	**41**	**45**	**49**	**53**	**57**	**61**	**65**	**69**
13	—	2	6	10	15	19	24	28	33	37	42	47	51	56	61	65	70	75	80	84
	—	**1**	**4**	**8**	**12**	**16**	**20**	**24**	**28**	**33**	**37**	**41**	**45**	**50**	**54**	**59**	**63**	**67**	**72**	**76**
14	—	2	7	11	16	21	26	31	36	41	46	51	56	61	66	71	77	82	87	92
	—	**1**	**5**	**9**	**13**	**17**	**22**	**26**	**31**	**36**	**40**	**45**	**50**	**55**	**59**	**64**	**69**	**74**	**78**	**83**
15	—	3	7	12	18	23	28	33	39	44	50	55	61	66	72	77	83	88	94	100
	—	**1**	**5**	**10**	**14**	**19**	**24**	**29**	**34**	**39**	**44**	**49**	**54**	**59**	**64**	**70**	**75**	**80**	**85**	**90**
16	—	3	8	14	19	25	30	36	42	48	54	60	65	71	77	83	89	95	101	107
	—	**1**	**6**	**11**	**15**	**21**	**26**	**31**	**37**	**42**	**47**	**53**	**59**	**64**	**70**	**75**	**81**	**86**	**92**	**98**
17	—	3	9	15	20	26	33	39	45	51	57	64	70	77	83	89	96	102	109	115
	—	**2**	**6**	**11**	**17**	**22**	**28**	**34**	**39**	**45**	**51**	**57**	**63**	**69**	**75**	**81**	**87**	**93**	**99**	**105**
18	—	4	9	16	22	28	35	41	48	55	61	68	75	82	88	95	102	109	116	123
	—	**2**	**7**	**12**	**18**	**24**	**30**	**36**	**42**	**48**	**55**	**61**	**67**	**74**	**80**	**86**	**93**	**99**	**106**	**112**
19	0	4	10	17	23	30	37	44	51	58	65	72	80	87	94	101	109	116	123	130
	—	**2**	**7**	**13**	**19**	**25**	**32**	**38**	**45**	**52**	**58**	**65**	**72**	**78**	**85**	**92**	**99**	**106**	**113**	**119**
20	0	4	11	18	25	32	39	47	54	62	69	77	84	92	100	107	115	123	130	138
	—	**2**	**8**	**13**	**20**	**27**	**34**	**41**	**48**	**55**	**62**	**69**	**76**	**83**	**90**	**98**	**105**	**112**	**119**	**127**

Wilcoxon Matched-Pairs Signed-Ranks Table

No. of pairs N	α levels for a one-tailed test .05	.025	.01	.005	N	α levels for a one-tailed test .05	.025	.01	.005
	α levels for a two-tailed test .10	.05	.02	.01		α levels for a two-tailed test .10	.05	.02	.01
5	0	—	—	—	28	130	116	101	91
6	2	0	—	—	29	140	126	110	100
7	3	2	0	—	30	151	137	120	109
8	5	3	1	0	31	163	147	130	118
9	8	5	3	1	32	175	159	140	128
10	10	8	5	3	33	187	170	151	138
11	13	10	7	5	34	200	182	162	148
12	17	13	9	7	35	213	195	173	159
13	21	17	12	9	36	227	208	185	171
14	25	21	15	12	37	241	221	198	182
15	30	25	19	15	38	256	235	211	194
16	35	29	23	19	39	271	249	224	207
17	41	34	27	23	40	286	264	238	220
18	47	40	32	27	41	302	279	252	233
19	53	46	37	32	42	319	294	266	247
20	60	52	43	37	43	336	310	281	261
21	67	58	49	42	44	353	327	296	276
22	75	65	55	48	45	371	343	312	291
23	83	73	62	54	46	389	361	328	307
24	91	81	69	61	47	407	378	345	322
25	100	89	76	68	48	426	396	362	339
26	110	98	84	75	49	446	415	379	355
27	119	107	92	83	50	466	434	397	373

* To be significant the T obtained from the data must be equal to or less than the value shown in the table.

SOURCE: From *Elementary Statistics*, Second Edition, by R. E. Kirk, Brooks/Cole Publishing, 1984.

Wilcoxon-Wilcox Multiple Comparisons Test Table

N	$\alpha = .05$ (two-tailed)							
	K = 3	K = 4	K = 5	K = 6	K = 7	K = 8	K = 9	K = 10
1	3.3	4.7	6.1	7.5	9.0	10.5	12.0	13.5
2	8.8	12.6	16.5	20.5	24.7	28.9	33.1	37.4
3	15.7	22.7	29.9	37.3	44.8	52.5	60.3	68.2
4	23.9	34.6	45.6	57.0	68.6	80.4	92.4	104.6
5	33.1	48.1	63.5	79.3	95.5	112.0	128.8	145.8
6	43.3	62.9	83.2	104.0	125.3	147.0	169.1	191.4
7	54.4	79.1	104.6	130.8	157.6	184.9	212.8	240.9
8	66.3	96.4	127.6	159.6	192.4	225.7	259.7	294.1
9	78.9	114.8	152.0	190.2	229.3	269.1	309.6	350.6
10	92.3	134.3	177.8	222.6	268.4	315.0	362.4	410.5
11	106.3	154.8	205.0	256.6	309.4	363.2	417.9	473.3
12	120.9	176.2	233.4	292.2	352.4	413.6	476.0	539.1
13	136.2	198.5	263.0	329.3	397.1	466.2	536.5	607.7
14	152.1	221.7	293.8	367.8	443.6	520.8	599.4	679.0
15	168.6	245.7	325.7	407.8	491.9	577.4	664.6	752.8
16	185.6	270.6	358.6	449.1	541.7	635.9	732.0	829.2
17	203.1	269.2	392.6	491.7	593.1	696.3	801.5	907.9
18	221.2	322.6	427.6	535.5	646.1	758.5	873.1	989.0
19	239.8	349.7	463.6	580.6	700.5	822.4	946.7	1072.4
20	258.8	377.6	500.5	626.9	756.4	888.1	1022.3	1158.1
21	278.4	406.1	538.4	674.4	813.7	955.4	1099.8	1245.9
22	298.4	435.3	577.2	723.0	872.3	1024.3	1179.1	1335.7
23	318.9	465.2	616.9	772.7	932.4	1094.8	1260.3	1427.7
24	339.8	495.8	657.4	823.5	993.7	1166.8	1343.2	1521.7
25	361.1	527.0	698.8	875.4	1056.3	1240.4	1427.9	1617.6

N	$\alpha = .01$ (two-tailed)							
	K = 3	K = 4	K = 5	K = 6	K = 7	K = 8	K = 9	K = 10
1	4.1	5.7	7.3	8.9	10.5	12.2	13.9	15.6
2	10.9	15.3	19.7	24.3	28.9	33.6	38.3	43.1
3	19.5	27.5	35.7	44.0	52.5	61.1	69.8	78.6
4	29.7	41.9	54.5	67.3	80.3	93.6	107.0	120.6
5	41.2	58.2	75.8	93.6	111.9	130.4	149.1	168.1
6	53.9	76.3	99.3	122.8	146.7	171.0	195.7	220.6
7	67.6	95.8	124.8	154.4	184.6	215.2	246.3	277.7
8	82.4	116.8	152.2	188.4	225.2	262.6	300.6	339.0
9	98.1	139.2	181.4	224.5	268.5	313.1	358.4	404.2
10	114.7	162.8	212.2	262.7	314.2	366.5	419.5	473.1
11	132.1	187.6	244.6	302.9	362.2	422.6	483.7	545.6
12	150.4	213.5	278.5	344.9	412.5	481.2	551.0	621.4
13	169.4	240.6	313.8	388.7	464.9	542.4	621.0	700.5
14	189.1	268.7	350.5	434.2	519.4	606.0	693.8	782.6
15	209.6	297.6	388.5	481.3	575.8	671.9	769.3	867.7
16	230.7	327.9	427.9	530.1	634.2	740.0	847.3	955.7
17	252.5	359.0	468.4	580.3	694.4	810.2	927.8	1046.5
18	275.0	391.0	510.2	632.1	756.4	882.6	1010.6	1140.0
19	298.1	423.8	553.1	685.4	820.1	957.0	1095.8	1236.2
20	321.8	457.6	597.2	740.0	885.7	1033.3	1183.3	1334.9
21	346.1	492.2	642.4	796.0	952.6	1111.6	1273.0	1436.0
22	371.0	527.6	688.7	853.4	1021.3	1191.8	1364.8	1539.7
23	396.4	563.8	736.0	912.1	1091.5	1273.8	1458.8	1645.7
24	422.4	600.9	784.4	972.1	1163.4	1357.6	1554.8	1754.0
25	449.0	638.7	833.8	1033.3	1236.7	1443.2	1652.8	1864.6

* To be significant the difference obtained from the data must be equal to or **larger than** the tabled value.

SOURCE: From *Some Rapid Approximate Statistical Procedures*, by F. Wilcoxon and R. Wilcox. Copyright © 1964 Lederle Laboratories, a division of American Cyanamid Co.

Spearman Table

No. of pairs	α levels			
	One-tailed test		Two-tailed test	
N	.01	.05	.01	.05
4	—	1.000	—	—
5	1.000	.900	—	1.000
6	.943	.829	1.000	.886
7	.893	.714	.929	.786
8	.833	.643	.881	.738
9	.783	.600	.833	.700
10	.745	.564	.794	.648
11	.709	.536	.755	.618
12	.678	.503	.727	.587
13	.648	.484	.703	.560
14	.626	.464	.679	.538
15	.604	.446	.654	.521
16	.582	.429	.635	.503

For samples larger than 16, use Table A.

* To be significant, the r_s obtained from the data must be equal to or larger than the value shown in the table.

SOURCE: From "Testing the Significance of Kendall's T and Spearman's r_s," by M. Nijsse, 1988, *Psychological Bulletin, 103*, 235–237. Copyright © 1988 American Psychological Association. Reprinted by permission.

APPENDIX B:
SPSS PRINTOUTS

Correlations: Hand Grip Strength and Forearm Muscle Size

Descriptive Statistics

	Mean	Std. Deviation	N
HANDGRIP	56.0000	15.52775	10
MUSCLE	27.0000	4.21637	10

Correlations

		HANDGRIP	MUSCLE
HANDGRIP	Pearson Correlation	1	.653*
	Sig. (2-tailed)	.	.040
	N	10	10
MUSCLE	Pearson Correlation	.653*	1
	Sig. (2-tailed)	.040	.
	N	10	10

*. Correlation is significant at the 0.05 level (2-tailed).

Correlations Sentence Years and Crime Severity

Descriptive Statistics

	Mean	Std. Deviation	N
YEARS	10.4000	6.14998	10
SEVERITY	5.5000	2.32140	10

Correlations

		YEARS	SEVERITY
YEARS	Pearson Correlation	1	.949**
	Sig. (2-tailed)	.	.000
	N	10	10
SEVERITY	Pearson Correlation	.949**	1
	Sig. (2-tailed)	.000	.
	N	10	10

**. Correlation is significant at the 0.01 level (2-tailed).

Regression: Hand Grip Strength and Forearm Muscle Size

Variables Entered/Removed[b]

Model	Variables Entered	Variables Removed	Method
1	HANDGRIP[a]	.	Enter

a. All requested variables entered.

b. Dependent Variable: MUSCLE

Model Summary

Model	R	R Square	Adjusted R Square	Std. Error of the Estimate
1	.653[a]	.427	.355	3.38551

a Predictors: (Constant), HANDGRIP

ANOVA[b]

Model		Sum of Squares	df	Mean Square	F	Sig.
1	Regression	68.306	1	68.306	5.960	.040[a]
	Residual	91.694	8	11.462		
	Total	160.000	9			

a. Predictors: (Constant), HANDGRIP

b. Dependent Variable: MUSCLE

Coefficients[a]

Model		Unstandardized Coefficients		Standardized Coefficients	t	Sig.
		B	Std. Error	Beta		
1	(Constant)	17.065	4.208		4.055	.004
	HANDGRIP	.177	.073	.653	2.441	.040

a. Dependent Variable: MUSCLE

Regression: Sentence Years and Crime Severity

Variables Entered/Removed[b]

Model	Variables Entered	Variables Removed	Method
1	YEARS[a]	.	Enter

a. All requested variables entered.

b. Dependent Variable: SEVERITY

Model Summary

Model	R	R Square	Adjusted R Square	Std. Error of the Estimate
1	.949[a]	.902	.889	.77257

a. Predictors: (Constant), YEARS

ANOVA[b]

Model		Sum of Squares	df	Mean Square	F	Sig.
1	Regression	43.725	1	43.725	73.257	.000[a]
	Residual	4.775	8	.597		
	Total	48.500	9			

a. Predictors: (Constant), YEARS

b. Dependent Variable: SEVERITY

Coefficients[a]

Model		Unstandardized Coefficients		Standardized Coefficients	t	Sig.
		B	Std. Error	Beta		
1	(Constant)	1.773	.499		3.550	.008
	YEARS	.358	.042	.949	8.559	.000

a. Dependent Variable: SEVERITY

T-Test: Room Color and Food Consumption

Group Statistics

	ROOM	N	Mean	Std. Deviation	Std. Error Mean
FOOD	BLACK	20	45.9500	3.33206	.74507
	YELLOW	20	50.0000	3.09499	.69206

Independent Samples Test

		Levene's Test for Equality of Variances		t-test for Equality of Means					95% Confidence Interval of the Difference	
		F	Sig.	t	df	Sig. (2-tailed)	Mean Difference	Std. Error Difference	Lower	Upper
FOOD	Equal variances assumed	.003	.955	-3.983	38	.000	-4.0500	1.01690	-6.10860	-1.99140
	Equal variances not assumed			-3.983	37.795	.000	-4.0500	1.01690	-6.10897	-1.99103

T-Test: Alzheimer Drug Comparison

Group Statistics

	DRUGGRP	N	Mean	Std. Deviation	Std. Error Mean
SYMP	PLACEBO	12	40.8333	3.04014	.87761
	DRUG	12	34.8333	5.62193	1.62291

Independent Samples Test

		Levene's Test for Equality of Variances		t-test for Equality of Means					95% Confidence Interval of the Difference	
		F	Sig.	t	df	Sig. (2-tailed)	Mean Difference	Std. Error Difference	Lower	Upper
SYMP	Equal variances assumed	6.471	.019	3.252	22	.004	6.0000	1.84500	2.17370	9.82630
	Equal variances not assumed			3.252	16.927	.005	6.0000	1.84500	2.10609	9.89391

T-Test: Religion Questionnaire

Paired Samples Statistics

		Mean	N	Std. Deviation	Std. Error Mean
Pair 1	PRETEST	13.6667	9	5.12348	1.70783
	POSTTEST	15.2222	9	5.71791	1.90597

Paired Samples Correlations

		N	Correlation	Sig.
Pair 1	PRETEST & POSTTEST	9	.796	.010

Paired Samples Test

		Paired Differences					t	df	Sig. (2-tailed)
					95% Confidence Interval of the Difference				
		Mean	Std. Deviation	Std. Error Mean	Lower	Upper			
Pair 1	PRETEST - POSTTEST	-1.5556	3.50397	1.16799	-4.2489	1.1378	-1.332	8	.220

T-Test: Sensitivity Training Workshop

Paired Samples Statistics

		Mean	N	Std. Deviation	Std. Error Mean
Pair 1	PRETEST	4.4000	5	2.07364	.92736
	POSTTEST	6.0000	5	2.23607	1.00000

Paired Samples Correlations

		N	Correlation	Sig.
Pair 1	PRETEST & POSTTEST	5	.970	.006

Paired Samples Test

		Paired Differences							
					95% Confidence Interval of the Difference				
		Mean	Std. Deviation	Std. Error Mean	Lower	Upper	t	df	Sig. (2-tailed)
Pair 1	PRETEST - POSTTEST	-1.6000	.54772	.24495	-2.2801	-.9199	-6.532	4	.003

Univariate Analysis of Variance: Happiness/Depression Drug

Between-Subjects Factors

		Value Label	N
DRUG	1.00	PLACEBO	10
	2.00	LOW DOSE	10
	3.00	HIGH DOSE	10

Descriptive Statistics

Dependent Variable: HAPPY

DRUG	Mean	Std. Deviation	N
PLACEBO	8.9000	3.03498	10
LOW DOSE	12.7000	4.32178	10
HIGH DOSE	16.2000	3.25918	10
Total	12.6000	4.59835	30

Tests of Between-Subjects Effects

Dependent Variable: HAPPY

Source	Type III Sum of Squares	df	Mean Square	F	Sig.
Corrected Model	266.600[a]	2	133.300	10.384	.000
Intercept	4762.800	1	4762.800	371.020	.000
DRUG	266.600	2	133.300	10.384	.000
Error	346.600	27	12.837		
Total	5376.000	30			
Corrected Total	613.200	29			

a. R Squared = .435 (Adjusted R Squared = .393)

Estimated Marginal Means

1. Grand Mean

Dependent Variable: HAPPY

		95% Confidence Interval	
Mean	Std. Error	Lower Bound	Upper Bound
12.600	.654	11.258	13.942

2. DRUG

Dependent Variable: HAPPY

DRUG	Mean	Std. Error	95% Confidence Interval	
			Lower Bound	Upper Bound
PLACEBO	8.900	1.133	6.575	11.225
LOW DOSE	12.700	1.133	10.375	15.025
HIGH DOSE	16.200	1.133	13.875	18.525

Post Hoc Tests

DRUG

Dependent Variable: HAPPY
Tukey HSD

(I) DRUG	(J) DRUG	Mean Difference (I-J)	Std. Error	Sig.	95% Confidence Interval Lower Bound	Upper Bound
PLACEBO	LOW DOSE	-3.8000	1.60231	.063	-7.7728	.1728
	HIGH DOSE	-7.3000*	1.60231	.000	-11.2728	-3.3272
LOW DOSE	PLACEBO	3.8000	1.60231	.063	-.1728	7.7728
	HIGH DOSE	-3.5000	1.60231	.092	-7.4728	.4728
HIGH DOSE	PLACEBO	7.3000*	1.60231	.000	3.3272	11.2728
	LOW DOSE	3.5000	1.60231	.092	-.4728	7.4728

Based on observed means
*. The mean difference is significant at the .05 level.

Homogeneous Subsets

HAPPY

Tukey HSD[a,b]

DRUG	N	Subset 1	2
PLACEBO	10	8.9000	
LOW DOSE	10	12.7000	12.7000
HIGH DOSE	10		16.2000
Sig.		.063	.092

Means for groups in homogeneous subsets are displayed.
Based on Type III Sum of Squares
The error term is Mean Square(Error) = 12.837.
  a. Uses Harmonic Mean Sample Size = 10.000.
  b. Alpha = .05

Profile Plots

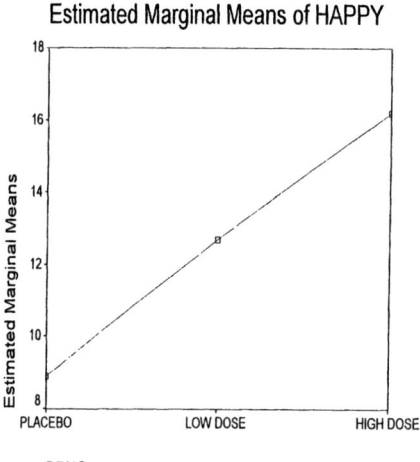

Estimated Marginal Means of HAPPY

Univariate Analysis of Variance: College Major and Motivation

Between-Subjects Factors

		Value Label	N
MAJOR	1.00	PSYCHOLOGY	8
	2.00	BUSINESS	8
	3.00	CHEMISTRY	8

Descriptive Statistics

Dependent Variable: MOTIVATE

MAJOR	Mean	Std. Deviation	N
PSYCHOLOGY	8.2500	1.38873	8
BUSINESS	9.0000	1.06904	8
CHEMISTRY	7.5000	1.77281	8
Total	8.2500	1.51083	24

Tests of Between-Subjects Effects

Dependent Variable: MOTIVATE

Source	Type III Sum of Squares	df	Mean Square	F	Sig.
Corrected Model	9.000[a]	2	4.500	2.172	.139
Intercept	1633.500	1	1633.500	788.586	.000
MAJOR	9.000	2	4.500	2.172	.139
Error	43.500	21	2.071		
Total	1686.000	24			
Corrected Total	52.500	23			

a. R Squared = .171 (Adjusted R Squared = .093)

Estimated Marginal Means

1. Grand Mean

Dependent Variable: MOTIVATE

Mean	Std. Error	95% Confidence Interval	
		Lower Bound	Upper Bound
8.250	.294	7.639	8.861

2. MAJOR

Dependent Variable: MOTIVATE

MAJOR	Mean	Std. Error	95% Confidence Interval	
			Lower Bound	Upper Bound
PSYCHOLOGY	8.250	.509	7.192	9.308
BUSINESS	9.000	.509	7.942	10.058
CHEMISTRY	7.500	.509	6.442	8.558

Post Hoc Tests

MAJOR

Dependent Variable: MOTIVATE
Tukey HSD

(I) MAJOR	(J) MAJOR	Mean Difference (I-J)	Std. Error	Sig.	95% Confidence Interval	
					Lower Bound	Upper Bound
PSYCHOLOGY	BUSINESS	-.7500	.71962	.559	-2.5639	1.0639
	CHEMISTRY	.7500	.71962	.559	-1.0639	2.5639
BUSINESS	PSYCHOLOGY	.7500	.71962	.559	-1.0639	2.5639
	CHEMISTRY	1.5000	.71962	.117	-.3139	3.3139
CHEMISTRY	PSYCHOLOGY	-.7500	.71962	.559	-2.5639	1.0639
	BUSINESS	-1.5000	.71962	.117	-3.3139	.3139

Based on observed means.

Homogeneous Subsets

MOTIVATE

Tukey HSD[a,b]

MAJOR	N	Subset 1
CHEMISTRY	8	7.5000
PSYCHOLOGY	8	8.2500
BUSINESS	8	9.0000
Sig.		.117

Means for groups in homogeneous subsets are displayed.
Based on Type III Sum of Squares
The error term is Mean Square(Error) = 2.071.

a. Uses Harmonic Mean Sample Size = 8.000

b. Alpha = .05

Profile Plots

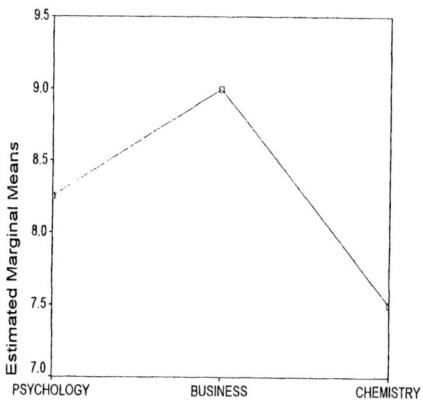

Estimated Marginal Means of MOTIVATE

General Linear Model: Soap Company Public Opinion

Within-Subjects Factors

Measure: MEASURE_1

OPINION	Dependent Variable
1	INITIAL
2	AFTSAMPL
3	SIXMONTH

Descriptive Statistics

	Mean	Std. Deviation	N
INITIAL	16.1429	3.89138	7
AFTSAMPL	32.5714	4.79086	7
SIXMONTH	25.2857	3.98808	7

Multivariate Tests[b]

Effect		Value	F	Hypothesis df	Error df	Sig.
OPINION	Pillai's Trace	.978	113.574[a]	2.000	5.000	.000
	Wilks' Lambda	.022	113.574[a]	2.000	5.000	.000
	Hotelling's Trace	45.430	113.574[a]	2.000	5.000	.000
	Roy's Largest Root	45.430	113.574[a]	2.000	5.000	.000

a. Exact statistic

b.
Design: Intercept
Within Subjects Design: OPINION

Mauchly's Test of Sphericity[b]

Measure: MEASURE_1

Within Subjects Effect	Mauchly's W	Approx. Chi-Square	df	Sig.
OPINION	.926	.386	2	.825

Tests the null hypothesis that the error covariance matrix of the orthonormalized transformed dependent variables is proportional to an identity matrix.

Mauchly's Test of Sphericity[b]

Measure: MEASURE_1

| Within Subjects Effect | Epsilon[a] | | |
	Greenhouse -Geisser	Huynh-Feldt	Lower-bound
OPINION	.931	1.000	.500

Tests the null hypothesis that the error covariance matrix of the orthonormalized transformed dependent variables is proportional to an identity matrix.

a. May be used to adjust the degrees of freedom for the averaged tests of significance. Corrected tests are displayed in the Tests of Within-Subjects Effects table.

b.

Design: Intercept
Within Subjects Design: OPINION

Tests of Within-Subjects Effects

Measure: MEASURE_1

Source		Type III Sum of Squares	df	Mean Square	F	Sig.
OPINION	Sphericity Assumed	948.667	2	474.333	99.279	.000
	Greenhouse-Geisser	948.667	1.862	509.552	99.279	.000
	Huynh-Feldt	948.667	2.000	474.333	99.279	.000
	Lower-bound	948.667	1.000	948.667	99.279	.000
Error(OPINION)	Sphericity Assumed	57.333	12	4.778		
	Greenhouse-Geisser	57.333	11.171	5.133		
	Huynh-Feldt	57.333	12.000	4.778		
	Lower-bound	57.333	6.000	9.556		

Tests of Within-Subjects Contrasts

Measure: MEASURE_1

Source	OPINION	Type III Sum of Squares	df	Mean Square	F	Sig.
OPINION	Linear	292.571	1	292.571	54.132	.000
	Quadratic	656.095	1	656.095	158.065	.000
Error(OPINION)	Linear	32.429	6	5.405		
	Quadratic	24.905	6	4.151		

Tests of Between-Subjects Effects

Measure: MEASURE_1
Transformed Variable: Average

Source	Type III Sum of Squares	df	Mean Square	F	Sig.
Intercept	12777.333	1	12777.333	287.490	.000
Error	266.667	6	44.444		

Estimated Marginal Means

1. Grand Mean

Measure: MEASURE_1

Mean	Std. Error	95% Confidence Interval	
		Lower Bound	Upper Bound
24.667	1.455	21.107	28.226

2. OPINION

Estimates

Measure: MEASURE_1

OPINION	Mean	Std. Error	95% Confidence Interval	
			Lower Bound	Upper Bound
1	16.143	1.471	12.544	19.742
2	32.571	1.811	28.141	37.002
3	25.286	1.507	21.597	28.974

Pairwise Comparisons

Measure: MEASURE_1

(I) OPINION	(J) OPINION	Mean Difference (I-J)	Std. Error	Sig.[a]	95% Confidence Interval for Difference[a]	
					Lower Bound	Upper Bound
1	2	-16.429*	.997	.000	-18.867	-13.990
	3	-9.143*	1.243	.000	-12.184	-6.102
2	1	16.429*	.997	.000	13.990	18.867
	3	7.286*	1.248	.001	4.232	10.340
3	1	9.143*	1.243	.000	6.102	12.184
	2	-7.286*	1.248	.001	-10.340	-4.232

Based on estimated marginal means

*. The mean difference is significant at the .05 level.

a. Adjustment for multiple comparisons: Least Significant Difference (equivalent to no adjustments).

Multivariate Tests

	Value	F	Hypothesis df	Error df	Sig.
Pillai's trace	.978	113.574[a]	2.000	5.000	.000
Wilks' lambda	.022	113.574[a]	2.000	5.000	.000
Hotelling's trace	45.430	113.574[a]	2.000	5.000	.000
Roy's largest root	45.430	113.574[a]	2.000	5.000	.000

Each F tests the multivariate effect of OPINION. These tests are based on the linearly independent pairwise comparisons among the estimated marginal means.

a. Exact statistic

Profile Plots

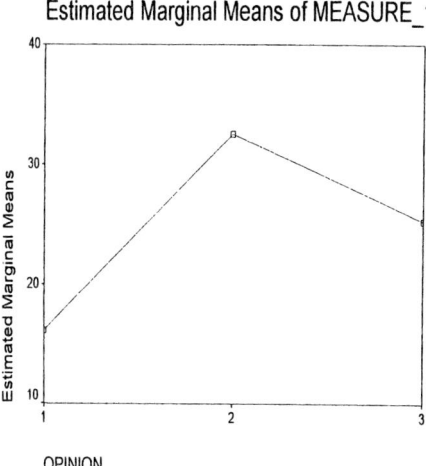

Estimated Marginal Means of MEASURE_1

OPINION

General Linear Model: Athletic Performance and Odors

Within-Subjects Factors

Measure: MEASURE_1

MOOD	Dependent Variable
1	PEPPERMT
2	JASMINE
3	LEMON

Descriptive Statistics

	Mean	Std. Deviation	N
PEPPERMT	8.6000	1.14018	5
JASMINE	6.2000	.83666	5
LEMON	6.4000	.54772	5

Multivariate Tests[b]

Effect		Value	F	Hypothesis df	Error df	Sig.
MOOD	Pillai's Trace	.904	14.182[a]	2.000	3.000	.030
	Wilks' Lambda	.096	14.182[a]	2.000	3.000	.030
	Hotelling's Trace	9.455	14.182[a]	2.000	3.000	.030
	Roy's Largest Root	9.455	14.182[a]	2.000	3.000	.030

a. Exact statistic

b.
Design: Intercept
Within Subjects Design: MOOD

Mauchly's Test of Sphericity[b]

Measure: MEASURE_1

Within Subjects Effect	Mauchly's W	Approx. Chi-Square	df	Sig.
MOOD	.645	1.318	2	.517

Tests the null hypothesis that the error covariance matrix of the orthonormalized transformed dependent variables is proportional to an identity matrix.

Mauchly's Test of Sphericity[b]

Measure: MEASURE_1

| Within Subjects Effect | Epsilon[a] | | |
	Greenhouse-Geisser	Huynh-Feldt	Lower-bound
MOOD	.738	1.000	.500

Tests the null hypothesis that the error covariance matrix of the orthonormalized transformed dependent variables is proportional to an identity matrix.

a May be used to adjust the degrees of freedom for the averaged tests of significance. Corrected tests are displayed in the Tests of Within-Subjects Effects table.

b

Design: Intercept
Within Subjects Design: MOOD

Tests of Within-Subjects Effects

Measure: MEASURE_1

Source		Type III Sum of Squares	df	Mean Square	F	Sig.
MOOD	Sphericity Assumed	17.733	2	8.867	16.625	.001
	Greenhouse-Geisser	17.733	1.476	12.018	16.625	.005
	Huynh-Feldt	17.733	2.000	8.867	16.625	.001
	Lower-bound	17.733	1.000	17.733	16.625	.015
Error(MOOD)	Sphericity Assumed	4.267	8	.533		
	Greenhouse-Geisser	4.267	5.902	.723		
	Huynh-Feldt	4.267	8.000	.533		
	Lower-bound	4.267	4.000	1.067		

Tests of Within-Subjects Contrasts

Measure: MEASURE_1

Source	MOOD	Type III Sum of Squares	df	Mean Square	F	Sig.
MOOD	Linear	12.100	1	12.100	14.235	.020
	Quadratic	5.633	1	5.633	26.000	.007
Error(MOOD)	Linear	3.400	4	.850		
	Quadratic	.867	4	.217		

Tests of Between-Subjects Effects

Measure: MEASURE_1
Transformed Variable: Average

Source	Type III Sum of Squares	df	Mean Square	F	Sig.
Intercept	749.067	1	749.067	607.351	.000
Error	4.933	4	1.233		

Estimated Marginal Means

1. Grand Mean

Measure: MEASURE_1

Mean	Std. Error	95% Confidence Interval	
		Lower Bound	Upper Bound
7.067	.287	6.271	7.863

2. MOOD

Estimates

Measure: MEASURE_1

MOOD	Mean	Std. Error	95% Confidence Interval	
			Lower Bound	Upper Bound
1	8.600	.510	7.184	10.016
2	6.200	.374	5.161	7.239
3	6.400	.245	5.720	7.080

Pairwise Comparisons

Measure: MEASURE_1

(I) MOOD	(J) MOOD	Mean Difference (I-J)	Std. Error	Sig.[a]	95% Confidence Interval for Difference[a]	
					Lower Bound	Upper Bound
1	2	2.400*	.400	.004	1.289	3.511
	3	2.200*	.583	.020	.581	3.819
2	1	-2.400*	.400	.004	-3.511	-1.289
	3	-.200	.374	.621	-1.239	.839
3	1	-2.200*	.583	.020	-3.819	-.581
	2	.200	.374	.621	-.839	1.239

Based on estimated marginal means

*. The mean difference is significant at the .05 level.

a. Adjustment for multiple comparisons: Least Significant Difference (equivalent to no adjustments).

Multivariate Tests

	Value	F	Hypothesis df	Error df	Sig.
Pillai's trace	.904	14.182[a]	2.000	3.000	.030
Wilks' lambda	.096	14.182[a]	2.000	3.000	.030
Hotelling's trace	9.455	14.182[a]	2.000	3.000	.030
Roy's largest root	9.455	14.182[a]	2.000	3.000	.030

Each F tests the multivariate effect of MOOD. These tests are based on the linearly independent pairwise comparisons among the estimated marginal means.

a Exact statistic

Profile Plots

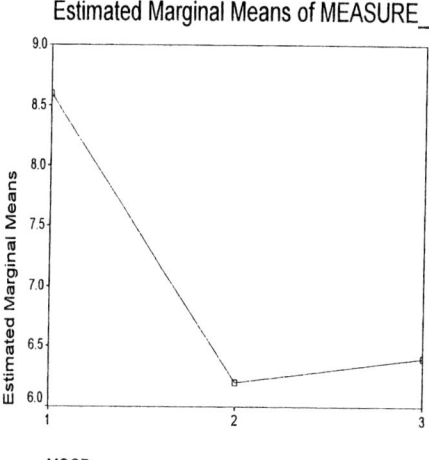

Estimated Marginal Means of MEASURE_1

MOOD

Univariate Analysis of Variance: State Dependent Learning

Between-Subjects Factors

		Value Label	N
LEARNING	1.00	DRUNK	16
	2.00	SOBER	16
RECALL	1.00	DRUNK	16
	2.00	SOBER	16

Descriptive Statistics

Dependent Variable: SCORE

LEARNING	RECALL	Mean	Std. Deviation	N
DRUNK	DRUNK	26.6250	1.92261	8
	SOBER	17.3750	2.38672	8
	Total	22.0000	5.21536	16
SOBER	DRUNK	18.1250	2.58775	8
	SOBER	31.1250	2.03101	8
	Total	24.6250	7.07931	16
Total	DRUNK	22.3750	4.91087	16
	SOBER	24.2500	7.41620	16
	Total	23.3125	6.26015	32

Tests of Between-Subjects Effects

Dependent Variable: SCORE

Source	Type III Sum of Squares	df	Mean Square	F	Sig.
Corrected Model	1073.375a	3	357.792	70.800	.000
Intercept	17391.125	1	17391.125	3441.353	.000
LEARNING	55.125	1	55.125	10.908	.003
RECALL	28.125	1	28.125	5.565	.026
LEARNING * RECALL	990.125	1	990.125	195.926	.000
Error	141.500	28	5.054		
Total	18606.000	32			
Corrected Total	1214.875	31			

a. R Squared = .884 (Adjusted R Squared = .871)

Estimated Marginal Means

1. Grand Mean

Dependent Variable: SCORE

Mean	Std. Error	95% Confidence Interval	
		Lower Bound	Upper Bound
23.313	.397	22.498	24.127

2. LEARNING

Dependent Variable: SCORE

LEARNING	Mean	Std. Error	95% Confidence Interval	
			Lower Bound	Upper Bound
DRUNK	22.000	.562	20.849	23.151
SOBER	24.625	.562	23.474	25.776

3. RECALL

Dependent Variable: SCORE

RECALL	Mean	Std. Error	95% Confidence Interval	
			Lower Bound	Upper Bound
DRUNK	22.375	.562	21.224	23.526
SOBER	24.250	.562	23.099	25.401

4. LEARNING * RECALL

Dependent Variable: SCORE

LEARNING	RECALL	Mean	Std. Error	95% Confidence Interval	
				Lower Bound	Upper Bound
DRUNK	DRUNK	26.625	.795	24.997	28.253
	SOBER	17.375	.795	15.747	19.003
SOBER	DRUNK	18.125	.795	16.497	19.753
	SOBER	31.125	.795	29.497	32.753

Profile Plots

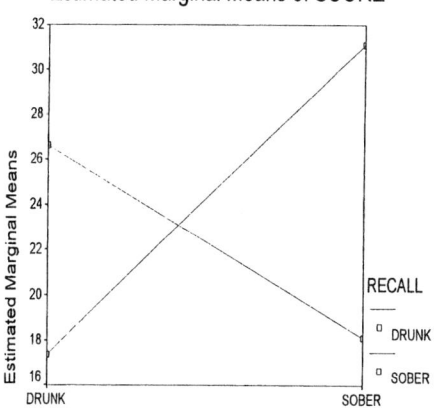

Estimated Marginal Means of SCORE

Univariate Analysis of Variance: Food, Experience and Mazes

Between-Subjects Factors

		Value Label	N
FOODTYPE	1.00	RAT CHOW	10
	2.00	SPECIAL BLEND	10
MAZEEXP	1.00	NO EXPERIENCE	10
	2.00	EXPERIENCE	10

Descriptive Statistics

Dependent Variable: MAZETIME

FOODTYPE	MAZEEXP	Mean	Std. Deviation	N
RAT CHOW	NO EXPERIENCE	14.4000	3.36155	5
	EXPERIENCE	13.2000	1.92354	5
	Total	13.8000	2.65832	10
SPECIAL BLEND	NO EXPERIENCE	13.6000	2.96648	5
	EXPERIENCE	7.2000	1.92354	5
	Total	10.4000	4.11501	10
Total	NO EXPERIENCE	14.0000	3.01846	10
	EXPERIENCE	10.2000	3.64539	10
	Total	12.1000	3.79612	20

Tests of Between-Subjects Effects

Dependent Variable: MAZETIME

Source	Type III Sum of Squares	df	Mean Square	F	Sig.
Corrected Model	163.800[a]	3	54.600	7.942	.002
Intercept	2928.200	1	2928.200	425.920	.000
FOODTYPE	57.800	1	57.800	8.407	.010
MAZEEXP	72.200	1	72.200	10.502	.005
FOODTYPE * MAZEEXP	33.800	1	33.800	4.916	.041
Error	110.000	16	6.875		
Total	3202.000	20			
Corrected Total	273.800	19			

a. R Squared = 598 (Adjusted R Squared = .523)

Estimated Marginal Means

1. Grand Mean

Dependent Variable: MAZETIME

Mean	Std. Error	95% Confidence Interval	
		Lower Bound	Upper Bound
12.100	.586	10.857	13.343

2. FOODTYPE

Dependent Variable: MAZETIME

FOODTYPE	Mean	Std. Error	95% Confidence Interval Lower Bound	95% Confidence Interval Upper Bound
RAT CHOW	13.800	.829	12.042	15.558
SPECIAL BLEND	10.400	.829	8.642	12.158

3. MAZEEXP

Dependent Variable: MAZETIME

MAZEEXP	Mean	Std. Error	95% Confidence Interval Lower Bound	95% Confidence Interval Upper Bound
NO EXPERIENCE	14.000	829	12.242	15.758
EXPERIENCE	10.200	.829	8.442	11.958

4. FOODTYPE * MAZEEXP

Dependent Variable: MAZETIME

FOODTYPE	MAZEEXP	Mean	Std. Error	95% Confidence Interval Lower Bound	95% Confidence Interval Upper Bound
RAT CHOW	NO EXPERIENCE	14.400	1.173	11.914	16.886
	EXPERIENCE	13.200	1.173	10.714	15.686
SPECIAL BLEND	NO EXPERIENCE	13.600	1.173	11.114	16.086
	EXPERIENCE	7.200	1.173	4.714	9.686

Profile Plots

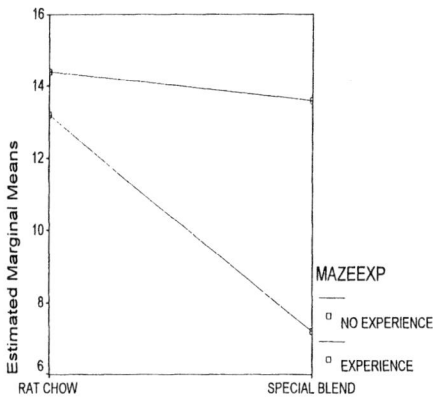

Estimated Marginal Means of MAZETIME